"十四五"职业教育国家规划教材

职业教育国家在线精品课程配套教材

高等职业教育**飞机电子设备维修**专业群新形态规划教材

民用航空

飞机维修手册应用

主 编 ♀ 乐 乐 耿明涛

副主编 ♀ 方 利 王志敏 吴德华 林 文

主 审 ♀ 王会来 魏 国

U0194577

中国水利水电出版社
www.waterpub.com.cn
·北京·

内 容 提 要

本教材为项目化教材，包括飞机发动机燃油渗漏分析、飞机尾白航行灯的安装、飞机防滞 / 自动刹车控制器故障隔离、飞机可收放着陆灯的故障分析、襟翼位置传感器线路故障的排除五大项目。配套双语工卡 31 个，涉及飞机空调系统、飞行控制系统、导航系统、灯光系统、燃油系统、灭火系统等民用航空器维修典型工作任务。

本教材通俗易懂，紧密联系民航维修实际，针对性强，可作为高等职业技术院校飞机维修相关专业的教学用书，也可作为 CCAR-147 培训机构的基本技能培训、全国"1+X"民用航空器航线维修职业技能等级培训及民航维修新进员工培训的参考教材，还可供飞机维修爱好者学习参考。

图书在版编目（C I P）数据

民用航空飞机维修手册应用 / 乐乐，耿明涛主编
. -- 北京 : 中国水利水电出版社，2021.9（2024.1 重印）
高等职业教育飞机电子设备维修专业群新形态规划教材
ISBN 978-7-5170-9980-2

Ⅰ. ①民… Ⅱ. ①乐… ②耿… Ⅲ. ①民用飞机－维修－高等职业教育－教材 Ⅳ. ①V267

中国版本图书馆CIP数据核字(2021)第188963号

策划编辑：周益丹　　　责任编辑：张玉玲　　　封面设计：梁　燕

书　名	高等职业教育飞机电子设备维修专业群新形态规划教材 **民用航空飞机维修手册应用** MINYONG HANGKONG FEIJI WEIXIU SHOUCE YINGYONG
作　者	主编　乐　乐　耿明涛 副主编　方　利　王志敏　吴德华　林　文 主审　王会来　魏　国
出版发行	中国水利水电出版社 （北京市海淀区玉渊潭南路 1 号 D 座　100038） 网址：www.waterpub.com.cn E-mail: mchannel@263.net（答疑） 　　　　sales@mwr.gov.cn 电话：（010）68545888（营销中心）、82562819（组稿）
经　售	北京科水图书销售有限公司 电话：（010）68545874、63202643 全国各地新华书店和相关出版物销售网点
排　版	北京万水电子信息有限公司
印　刷	三河市德贤弘印务有限公司
规　格	184mm×260mm　16 开本　16.5 印张　361 千字
版　次	2021 年 9 月第 1 版　2024 年 1 月第 4 次印刷
印　数	6001—10000 册
定　价	59.00 元

前　言

　　民用航空飞机维修手册是民用航空维护人员从事民用航空器维护工作，学习民用航空飞机相关系统运转状况，掌握飞机基本结构，了解民用航空器基本维护知识，掌握故障隔离方法，保证飞机持续适航性等的重要依据。近年来，我国航空科技的发展日新月异，具有自主知识产权的国产大飞机 C919 顺利取得型号合格证，即将投入运营，对以欧美飞机持续适航维护为基础的飞机维护体系提出了挑战。随着我国民航事业的迅猛发展，中国民航（CAAC）与 FAA、EASA 适航体系匹配度日益增强，民航维修标准国际化程度也愈趋提高，国内大部分航空公司均使用全英文机务维修工卡，业内对民航机务维修人员全英文手册查询、阅读、应用的能力提出了更高的要求。

　　培养高素质民航机务维修人才，高质量的民用航空飞机维修手册应用教材不可或缺。遗憾的是，目前市面上与民航机务维修岗位零对接且注重民航机务维修人员职业品格培养的教材不多。纵观业内本课程教材市场，大致分为两类。一是已公开出版的教材，其大部分偏重手册本身结构与功能的介绍，与民航机务修理任务关联性不大，对飞机各系统的背景认识不深，缺乏常用飞机维修手册应用以及多种手册综合应用的系统训练，不能完全满足民航维修岗位对手册综合应用能力的需求；二是院校自编的校本教材或讲义，尽管此类教材针对性较强，但不够系统和全面，民用航空器维修典型案例的分析不充分，且无配套项目汉英双语工卡及相应的实操练习。同时大部分教材文字描述过多，图例刻板；手册应用所涉及的空调、通信、电源、导航等诸多专业知识相对抽象，仅通过文字描述及二维图画，学生难以快速、准确理解系统的相关内容。

　　本教材正是针对目前市面上教材存在的缺陷，基于培养高素质民航机务维修人员的急迫需要而编写的。教材以民航机务维修人员工作任务和工作过程为逻辑起点，依据中国民航法规、民用航空器维修人员执照基础考试大纲、民用航空器维修基础培训大纲，结合民用航空器维修典型的修理任务构建项目式框架；紧扣从"手册认知、规范查询、熟练应用到综合应用"的实践主线，强调了各类手册"How to use"；采用案例引导，学习者跟随查询的方式，结合微课、动画等教学资源，辅助教师实施以行动为导向的教学，培养学习者主动参与构建，能够胜任执行完整的修理任务的职业行动能力及职业品格；语言通俗易懂，做到了集专业性、科普性和实用性于一体。本教材具有如下鲜明特色：

　　1. 以民航企业真实案例为载体，重组重构项目内容

　　本教材紧密联系民航维修实际，以岗位职业能力为导向，将 AMM、IPC、FIM、

SSM、WDM 等波音飞机常用维修手册的功能、结构、查询、应用及职业品格等教学目标，以民用航空器维修典型案例为载体，重组了飞机发动机燃油渗漏分析（AMM）、飞机尾白航行灯的安装（IPC、AMM）、飞机防滞 / 自动刹车控制器故障隔离（FIM、AMM）、飞机可收放着陆灯的故障分析（SSM、FIM、AMM）、襟翼位置传感器线路故障的排除（FIM、AMM、WDM）五大项目。每个项目的手册各有侧重，随着项目递进，其所涉及的手册也在不断叠加，更加真实地还原了手册实际应用情况，有利于培养学习者综合性项目的解决能力。

2. 以学习者自由选择为中心，编写差异化双语工卡

本教材紧跟民航维修标准国际化趋势，紧贴业内对民航机务维修人员全英文手册查询、阅读、应用能力的基本要求，遵守民航机务维修手册中规定的工作前、工作中、工作后的步骤，以及工作程序和注意事项等，配套编写了 AV/ME 汉英双语机务维修工卡共 31 个，设置 I（难）、II（中）、III（易）三个难度等级，涵盖飞机维护手册（AMM）、零部件图解目录手册（IPC）、故障隔离手册（FIM）、系统原理图手册（SSM）、线路图手册（WDM）等手册的查询与应用，可供不同专业方向、不同学情的学习者自由选择，亦便于教师开展差异化教学。

3. 以民用航空器维修过程为序，细化细分工作任务

严格按照相关手册的内容工作是机务人员必须遵守的规章制度，稍有疏忽遗漏的环节就有可能造成严重的不良后果。工卡中设置的工作前、工作中、工作后的步骤，以及工作程序和注意事项，手册中都有明确指示和相应要求。本教材以民用航空器维修过程为序，细化细分了若干任务；任务之间存在着递进关系，不可调换任务顺序；任务完成，则相应项目完成。例如：发现故障，通过 FIM 手册找到相应的故障隔离程序；按照程序要求对应 AMM 程序做相应测试来隔离故障；通过 IPC 查找相应需更换硬件；根据 SSM 系统图做故障判断；最后需依照 WDM 完成线路故障查找及相关修理工作。

4. 以立德树人为根本任务，有机融入思政元素

坚持为党育人，为国育才。全面贯彻党的教育方针，认真落实立德树人根本任务，教材以"敬航空、爱航空、懂航修"为主线，从国家、行业、个人三个层次，针对国家民航战略规划、国产大飞机现状与未来发展、民航企业及维修岗位规范、民航维修机务人员个人发展等方面深度挖掘思政元素，创建"蓝天魂•三敬情•工匠心"课程思政体系，将民航人的责任与使命，民航局"三严"要求，民航机务维修人"三敬畏"，及对隐患"零容忍"、对疑问"零保留"、"四个意识"、"五个到位"等职业素养有机融入到教学内容中，潜移默化地培养学生爱岗敬业劳动态度和精益求精的工匠精神。

5. 以多样化配套资源为支撑，全方位辅助教学

本教材是高等职业教育航空类新形态活页式教材。所有重要的知识点和技能点均配有图片、微课、动画、虚拟仿真、习题等丰富的数字化资源。其中，微课类资源学习者可通过手机扫描书中的二维码，登录万水云课堂在线观看；也可登录中国大学慕课网

页（或手机下载中国大学慕课 APP），搜索课程"飞机维修手册与文件的使用"，报名线上学习。本教材既可作为高等职业技术院校飞机维修相关专业的教学用书，也可作为 CCAR-147 培训机构的基本技能培训及民航维修新进员工培训的参考教材，还可供飞机维修爱好者学习参考。授课教师如需要本书配套的教学课件资源，可发送邮件至邮箱 17121245@qq.com 索取。

6. 以活页式结构呈现为表征，保持内容动态开放

本教材作为新型活页式教材，形式上更加新颖活泼，内容上保持动态开放，可主动适应新机型、CCAR-66R3 民用航空器维修执照新标准、波音飞机维修手册查询系统 TOOLBOX 新技术等快速迭代，尤其是国产大飞机未来进入维修阶段后，其维修手册以及典型工作案例可迅速补充进教材；可随时根据 AV、ME 不同专业学习者需要进行拆卸和二次组装，满足民航机务维修岗位不同群体的个性化细分需求；可支持民航机务维修学习者在不破坏装订外形的情况下，取出或加入内容，调整模块顺序、自行整理组合项目，如交作业、夹入笔记、替换旧内容、加入企业学习内容等。

本教材由长沙航空职业技术学院乐乐教授、东方航空有限公司西北分公司资深维修工程师耿明涛先生担任主编，中国人民解放军第五七〇六工厂方利、长沙航空职业技术学院王志敏副教授、吴德华教授、林文副教授担任副主编，赵迎春副教授负责工卡英文翻译，薛笑雷、司维钊、李昭飞老师参与了资源录制，是校企深度合作的集体智慧的体现。本教材由 Ameco 航空技工学校校长、民用航空器系列教材主编王会来先生和中国民航大学副教授、全国职业院校技能大赛高职组"飞机发动机拆装调试与维修"赛项（线标组）裁判长魏国先生主审。在审稿过程中，他们为本书的编写提出了许多建设性意见，在此表示衷心感谢。另外，衷心感谢原中国民用航空器维修人员执照考试管理中心主任、中国民航大学任仁良教授对本书提出的宝贵建议，湖南省芙蓉教学名师、湖南省高职思政教育教学团队负责人、长沙航空职业技术学院雷世平教授给予的课程思政指导，中国南方航空股份有限公司湖南分公司飞机维修厂质量管理室主任麻湘先生、深圳航空有限责任公司资深培训工程师陈明先生给予的技术指导，长沙航空职业技术学院主管教学副院长朱国军、科研处处长彭圣文、电子学院院长易江义、CCAR-147 机务培训中心质量经理江游对教材给予的大力支持。感谢飞机电子设备维修专业教研室的同事以及提供过帮助的学生。最后，向所有为本书提供大量参考资料、各种型号常用飞机维修手册和帮助的机务维修一线的领导及工作人员表示诚挚的感谢。

限于编者的知识水平和实践经验，本教材难免存在错漏和不妥之处，恳请读者、同行批评指正，以便于本教材在今后修订过程中的改进。

编 者

2022 年 11 月

目 录

项目 3　FIM 等手册应用：飞机防滞 /
自动刹车控制器故障隔离

项目 4　SSM 等手册应用：飞机可收放
着陆灯的故障分析

项目 5　WDM 等手册应用：襟翼位置传感器线路故障的排除

工卡索引

资源索引

项目 1

AMM 手册应用：飞机发动机燃油渗漏分析
Application of AMM: Analysis of Fuel Leakage of Aircraft Engine

【大国工匠】航空发动机首席技师
李志强：飞天梦想，指尖传奇

项目导读

　　近年来在世界各地的主要航空公司，不管是波音还是空客公司的飞机执行冬季航班时都报告了多起燃油泄漏故障。某年冬季，某航空波音 737-NG 型飞机降落广州白云机场，机务人员在执行飞行任务前发现 CFM56-7B 型发动机底部排油桅杆漏油。这时候机务人员在确认渗漏的油液是燃油并确定燃油的渗漏情况后，立刻将发动机的漏油情况通报了基地维修控制中心，维修人员通过查找飞机维护手册（AMM）找到漏油标准，发现飞机静态燃油渗漏是 26 滴 / 分钟，未超标，飞机正常放行。这样，在确保飞机安全飞行的前提下，同时保障了航班的正常运行，提高了航空公司的运营效率。那么，飞机发动机燃油渗漏值在什么范围内可以正常放行？达多少值需要及时维护？

教学目标

★掌握飞机维护手册（AMM）的功能。
★掌握飞机维护手册（AMM）的结构。
★掌握飞机维护手册（AMM）的查询方法。
★掌握飞机维护方案的制定。

学习导航

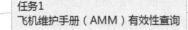

项目1 AMM 手册应用：飞机发动机燃油渗漏分析

任务1 飞机维护手册（AMM）有效性查询
1. 飞机维护手册（AMM）的功能与内容
2. 飞机维护手册（AMM）前言的编排
3. 飞机维护手册的有效性

任务2 飞机发动机燃油渗漏章节查询
1. 《ATA 100 规范》
2. 《ATA 100 规范》的章节编码规则
3. SDS 的功能与结构
4. 飞机发动机燃油渗漏故障

任务3 飞机发动机燃油渗漏分析
1. PP 手册的功能与结构
2. PP 手册页码段的划分

任务 1 飞机维护手册（AMM）有效性查询

任务描述

飞机发动机燃油渗漏故障维护需使用飞机维护手册，查询飞机发动机燃油渗漏标准来明确该故障的维护程序。飞机维护手册（Aircraft Maintenance Manual，AMM）是外场维护使用频率很高的手册，是飞机工作人员的工作指南。AMM 手册是客户化手册，在它的封面上明确地标出该手册是属于某个客户的，该手册只能在该客户的范围内针对某型飞机使用，而对其他客户相同型号的飞机是无效的。

如何查询飞机的有效性呢？飞机维护手册的有效性列表中能查到哪些重要信息呢？

任务要求

- 了解飞机维护手册飞机有效性列表。
- 了解客户化手册的特点。

知识链接

AMM 能干啥？

1. 飞机维护手册（AMM）的功能与内容

中国民用航空规章规定飞机的维护和修理内容，修理工艺、程序等必须符合法定技术文件的要求。其中，AMM 是飞机维修单位对飞机进行维护和修理时非常重要的手册。AMM 由飞机制造厂商发布，依据各种组件、系统、APU、发动机的制造厂商、供货商提供的资料和技术数据综合编写而成。AMM 为飞机维护和维修提供了系统和零部件的说明，以及各种勤务、维护、检查、排除故障、系统功能试验、调节、清洁、修理和更换零部件等工作的详细技术标准和工艺程序资料等。这些技术资料为飞机航线维护和机库内维护（修）飞机提供技术指导和技术支持，为制定飞机维护和修理方案提供依据。

在维修文件历史的传承中，出现了很多维护手册内容的分支，因此不同时代出现了不同内容的维护手册，新旧不同版本的维护手册的内容也不尽相同。波音 737-600/700/800/900 飞机的维护手册在内容的分类上，将通用性、原理性的信息组成 Part I 部分，称为系统描述部分（Systems Description Section，SDS），它继承了原来（波音 737-300/400/500 飞机）在 001 ～ 099 页部分的概述内容，由于这部分内容的错误不涉及工作内容，波音公司可以免责。将维护工作部分组成 Part II 部分，称为施工与程序（Practices and Procedures，PP）。从某种意义上讲，AMM 手册就是所有针对航线可更换件（Line Replaceable Unit，LRU）进行的维护步骤和程序的集合。

飞机维护手册主要包括：

（1）对飞机各系统以及部件的描述，如对液压、燃油、操纵系统和电器等系统以及起落架等部件的构成、功能、位置等说明。

（2）系统或者部件故障诊断，包含系统或者部件常见的故障、引起故障的原因、故障诊断程序以及推荐的修理。

（3）各种勤务工作，包括添加燃油、滑油和机轮轮胎充气等程序。

（4）维修实践，包括常规修理工作程序、简要的拆装、调节和检测等程序。

（5）拆卸与安装，包括拆装的准备工作、所需的设备、材料以及零部件详细的拆装程序等。

（6）调节与测试，包括调节程序和测试程序，测试程序包含部件功能检测和系统检测。

（7）检测与检查，为航线维护提供各种磨损极限、零部件的裂纹、凹坑和间隙以及液面指示、压力和电阻等数据与标准。

（8）清洗与喷涂，包括恢复、修理零部件各种表面的程序和使用化学溶剂的注意事项等。

（9）已批准的修理，包括塑料表面、密封件、螺栓孔等细小的损伤修理，但不包括结构（件）的修理。若要对结构进行修理必须按飞机结构修理手册的规定去修理。

图 1-1　前言编排

2. 飞机维护手册（AMM）前言的编排

AMM 的前言包含标题页（Title）、飞机有效性对照表（Effective Aircraft）、手册发送说明（Transmittal Letter）、修订重点说明（Highlights）、有效页清单（Effective Pages）、章节有效性（Effective Chapters）、修订记录（Revision Record）、临时修订记录（Record of Temporary Revision）、维修服务通告清单（Service Bulletin List）、简介（Introduction）等。前言编排如图 1-1 所示。

（1）标题页（Title）：手册的版权页，如图 1-2 所示。

图 1-2　标题页

（2）飞机有效性对照表（Effective Aircraft）：显示适用于该手册的所有飞机的各种编号，如图 1-3 所示。

BOEING　　　　　737-600/700/800/900 AIRCRAFT MAINTENANCE MANUAL

This manual is applicable to the aircraft in this list:

| Model-Series | Operator | | Manufacturer | | | Registration Number |
	Identification Code	Effectivity Code	Block Number	Serial Number	Line Number	
737-7W0	XXX	001	YA811	29912	140	B-2639
737-7W0	XXX	002	YA812	29913	148	B-2640
737-7W0	XXX	003	YA813	30074	292	B-2503
737-7W0	XXX	004	YA814	30075	311	B-2502

EFFECTIVE AIRCRAFT

图 1-3　飞机有效性对照表

（3）手册发送说明（Transmittal Letter）：致手册持有者的信函，如图 1-4 所示。

BOEING　　　　　737-600/700/800/900 AIRCRAFT MAINTENANCE MANUAL

XXX XXX XXX Airlines
XXX
Revision No. 29
Feb 10/2006

To: All holders of this Boeing Document D633A101- XXX

Attached is the current revision to the Part I Boeing 737-600/700/800/900 Aircraft Maintenance Manual.

The Aircraft Maintenance Manual (SDS) is furnished either as a printed manual, on microfilm, or digital products, or any combination of the three. This revision replaces all previous microfilm cartridges or digital products. All microfilm and digital products are reissued with all obsolete data deleted and all updated pages added.

For printed manuals, changes are indicated on the List of Effective Pages (LEP). The pages which are revised will be identified on the LEP by an R (Revised), A (Added), O (Overflow, i.e. changes to the document structure and/or page layout), or D (Deleted). Each page in the LEP is identified by Chapter-Section-Subject number, page number and page date.

Pages replaced or made obsolete by this revision should be removed and destroyed.

ATTENTION

IF YOU RECEIVE PRINTED REVISIONS, PLEASE VERIFY THAT YOU HAVE RECEIVED AND FILED THE PREVIOUS REVISION. BOEING MUST BE NOTIFIED WITHIN 30 DAYS IF YOU HAVE NOT RECEIVED THE PREVIOUS REVISION. REQUESTS FOR REVISIONS OTHER THAN THE PREVIOUS REVISION WILL REQUIRE A COMPLETE MANUAL REPRINT SUBJECT TO REPRINT CHARGES SHOWN IN THE DATA AND SERVICES CATALOG.

TRANSMITTAL LETTER

图 1-4　手册发送说明

（4）修订重点说明（Highlights）：列出了手册换版时各章节的更改位置和更改原因，如图 1-5 所示。

BOEING ® 737-600/700/800/900 AIRCRAFT MAINTENANCE MANUAL

Location of Change	Description of Change	Location of Change	Description of Change
CHAPTER 22		PAGESET 22-11-00-096	Added the changes that show the effect of the installation of the Global Landing System (GLS) on the DFCS.
22-11-00			
PAGESET 22-11-00-025	Added the changes that show the effect of the installation of the Global Landing System (GLS) on the DFCS.		
		CHAPTER 23	
		23-11-00	
PAGESET 22-11-00-026	Added the changes that show the effect of the installation of the Global Landing System (GLS) on the DFCS.	PAGESET 23-11-00-005	Changed the safe distance for HF transmissions to ten feet.
		PAGESET 23-11-00-008	Changed the HF Transceiver SQL/LAMP TEST button text to TEST button.
PAGESET 22-11-00-030	Added the changes that show the effect of the installation of the Global Landing System (GLS) on the DFCS.		Changed the SQL/LAMP Test Button text to Test Button and changed the data.
PAGESET 22-11-00-061	Added the changes that show the effect of the installation of the Global Landing System (GLS) on the DFCS.	PAGESET 23-11-00-010	Changed the safe distance for HF transmissions to ten feet.
		CHAPTER 24	
		24-20-00	
PAGESET 22-11-00-076	Added the changes that show the effect of the installation of the Global Landing System (GLS) on the DFCS.	PAGESET 24-20-00-008	Changed the power distribution panel 2 light name and condition.
PAGESET 22-11-00-094	Added data for Integrated Approach Navigation mode.	**CHAPTER 25**	
		25-60-00	
	Added the changes that show the effect of the installation of the Global Landing System (GLS) on the DFCS.	PAGESET 25-60-00-006	Added the data for the breathing equipment installation.

HIGHLIGHTS

图 1-5　修订重点说明

（5）有效页清单（Effective Pages）：查验章节页是否现行有效，如图 1-6 所示。

BOEING ® 737-600/700/800/900 AIRCRAFT MAINTENANCE MANUAL

Subject/Page	Date	COC	Subject/Page	Date	COC
TITLE PAGE			RECORD OF TEMPORARY REVISIONS		
O 1	Feb 10/2006		1	Oct 10/2002	
2	BLANK		2	Oct 10/2002	
EFFECTIVE AIRCRAFT			SERVICE BULLETIN LIST		
1	Oct 10/2003		1	Oct 10/2003	
2	BLANK		2	BLANK	
TRANSMITTAL LETTER			INTRODUCTION		
O 1	Feb 10/2006		1	Feb 10/2005	
2	Oct 10/2002		2	Feb 10/2005	
HIGHLIGHTS			3	Feb 10/2003	
O 1	Feb 10/2006		4	Oct 10/2002	
O 2	Feb 10/2006		5	Feb 10/2003	
O 3	Feb 10/2006		6	Oct 10/2002	
O 4	Feb 10/2006		7	Oct 10/2002	
EFFECTIVE PAGES			8	Jun 10/2003	
1	Feb 10/2006		9	Oct 10/2002	
2	BLANK		10	BLANK	
EFFECTIVE CHAPTERS					
O 1	Feb 10/2006				
O 2	Feb 10/2006				
REVISION RECORD					
1	Oct 10/2002				
2	Oct 10/2002				

A = Added, R = Revised, D = Deleted, O = Overflow, C = Customer Originated Change

EFFECTIVE PAGES

图 1-6　有效页清单

其中：A 表示增加页；R 表示改版页；D 表示删除页；O 表示覆盖页；C 表示应客户要求发起的更改。

（6）章节有效性（Effective Chapters）：描述了各章节增加、改动的信息，如图 1-7 所示。

（7）修订记录（Revision Record）：用来记录每一次定期修订。飞机维护手册有正常修订和临时修订两种。其中，B737 飞机维护手册每年有三次正常修订服务，日期是 2 月 10 日、6 月 10 日和 10 月 10 日。对修改过的节或页面将在有效页清单上用 R（已修改）、A（已增加）或 D（已删除）来标识。如图 1-8 所示为定期修订记录，即为正常修订。

BOEING 737-600/700/800/900 AIRCRAFT MAINTENANCE MANUAL

	Chapter	Date	Title
	21	Oct 10/2005	Air Conditioning
R	22	Feb 10/2006	Autoflight
R	23	Feb 10/2006	Communications
R	24	Feb 10/2006	Electrical Power
R	25	Feb 10/2006	Equipment and Furnishings
	26	Oct 10/2003	Fire Protection
R	27	Feb 10/2006	Flight Controls
	28	Oct 10/2005	Fuel
R	29	Feb 10/2006	Hydraulic Power
R	30	Feb 10/2006	Ice and Rain Protection
R	31	Feb 10/2006	Indicating and Recording Systems
	32	Oct 10/2005	Landing Gear
R	33	Feb 10/2006	Lights
R	34	Feb 10/2006	Navigation
R	35	Feb 10/2006	Oxygen
R	36	Feb 10/2006	Pneumatic
	38	Oct 10/2005	Water and Waste
R	49	Feb 10/2006	Auxiliary Power System
	51	Jun 10/2002	Structures
	52	Feb 10/2004	Doors
	53	Jun 10/2002	Fuselage
	54	Oct 10/2005	Nacelles and Pylons
	55	Jun 10/2002	Stabilizers
	56	Oct 10/2005	Windows
	57	Oct 10/2004	Wings

A = Added, R = Revised

EFFECTIVE CHAPTERS

图 1-7　章节有效性

BOEING 737-600/700/800/900 AIRCRAFT MAINTENANCE MANUAL

All revisions to this manual will be accompanied by transmittal sheet bearing the revision number. Enter the revision number in numerical order, together with the revision date, the date filed and the initials of the person filing.

Revision		Filed		Revision		Filed	
Number	Date	Date	Initials	Number	Date	Date	Initials

REVISION RECORD

图 1-8　定期修订记录

另外，在飞机有效性清单页上修改过的地方在其左边空白处用修改竖杠表示。手册中除封面和每章的封面外，其每一页的右下角都标有该页编出的日期，同时在页脚中间标识用户文件编号，有效性清单（LEP）中记录这些信息用来作为手册内容的权限。

（8）临时修订记录（Record of Temporary Revision）：用来记录每一次临时修订。在手册两次连续正式修订期间，如要对手册内容进行修订则进行临时修订。每次临时修订只能修订一个项目的内容，修订后也应做好相应记录（图1-9）。

RECORD OF TEMPORARY REVISION

图 1-9　临时修订记录

（9）维修服务通告清单（Service Bulletin List）：包括服务通告号、服务通告涉及的ATA 章、通告状态、出版日期及其对手册有效性的影响等，如图1-10 所示。

在 Started/Completed 项目中，如果标 S，则说明这份服务通告尚未完成，如果标 C，则说明这份服务通告已经完成。

（10）简介（Introduction）：描述了手册的结构章节编排规则、有效性识别、页段含义、手册使用方法、任务编号系统，还包括消耗材料清单、工具清单、消耗材料和工具供应商清单，如图1-11 所示。

3. 飞机维护手册的有效性

有效性是连接维修文件和机务维修实践活动的纽带，是维修文件在适航的要求中，适用于指定飞机以及飞机机载设备的依据。在有效性未知的情况下，任何维修文件不得使用在任何航空器上。

（1）飞机维护手册的有效性。飞机维护手册（AMM）的有效性是指该手册对某一特定机型的哪些飞机有效。以 B737-700 型飞机的 AMM 为例，在手册的扉页飞机有效性对照表（Effective Aircraft）中标明了该手册适用于哪些 B737-700 的飞机，如图1-3 所示。

737-600/700/800/900 AIRCRAFT MAINTENANCE MANUAL

Number	Incorporated	Started/ Completed	ATA	Subject
-	No Effect		-	-

A = Added, R = Revised

SERVICE BULLETIN LIST

图 1-10   维修服务通告清单

737-600/700/800/900 AIRCRAFT MAINTENANCE MANUAL

INTRODUCTION

General

This Publication is Part I of the 737-600/700/800/900 MAINTENANCE MANUAL, the Systems Description Section (SDS). It has been prepared by the Boeing Commercial Airplane Group in accordance with Air Transport Association of America Specification No. 100, section 2-17, Specification for Manufacturers' Technical Data. It contains information on component location, system operation, and Training Information Points for all systems and equipment installed in the 737-6/7/8/9 airplanes listed herein normally requiring such action on the line or in the maintenance hangar. Information required to check, repair, adjust, or test units or assemblies, normally performed away from the airplane because of the need for special equipment, is contained in the Boeing 737 Component Maintenance Manual or vendor's component maintenance manual(s).

NOTE: THIS MANUAL WAS PREPARED SPECIFICALLY TO COVER THE BOEING AIRPLANES LISTED IN THE LIST OF EFFECTIVE AIRPLANES, FOUND AFTER THE TITLE PAGE OF THIS DOCUMENT. IT CONTAINS INSTRUCTIONS AND INFORMATION APPLICABLE SOLELY TO THOSE SPECIFIC AIRPLANES.

NOTE: OPERATORS ARE SOLELY RESPONSIBLE FOR 1) THE ACCURACY AND VALIDITY OF ALL INFORMATION FURNISHED BY THE OPERATOR OR ANY OTHER PARTY BESIDES BOEING: AND 2) ENSURING THE MAINTENANCE DOCUMENTATION THEY ARE USING IS COMPLETE AND MATCHES THE CURRENT CONFIGURATION OF THE AIRPLANE, AND FOR THE OPERATORS RECEIVING ACTIVE REVISION SERVICE, THAT ANY MODIFICATIONS TO THE AIRPLANE ARE PROPERLY REFLECTED IN THE MAINTENANCE INSTRUCTIONS CONTAINED IN THIS MANUAL. THE BOEING COMPANY ASSUMES NO RESPONSIBILITY IN THIS REGARD.

NOTE: THIS MANUAL IS NOT SUITABLE FOR USE, INCLUDING WITHOUT LIMITATION, GENERAL INSTRUCTIONS OR TRAINING, FOR ANY AIRPLANES NOT LISTED HEREIN, NOR DOES IT NECESSARILY APPLY TO LISTED AIRPLANES THAT HAVE BEEN CONVEYED TO OTHER OPERATORS.

INTRODUCTION

图 1-11   简介

（2）正文内容页的修订有效性。正文的每一章最前面都有飞机有效页清单（Effective Pages），正文内容页的修订有效性是指手册中每页对应的有效性，通过核对内页右下角的修订日期和有效页清单中给出的修订日期，判断查阅手册具体章节页是否现行有效，如图 1-2 所示。

（3）手册正文页的飞机构型有效性。手册正文页的飞机构型有效性在手册正文页的左下角用文字表示。如果某页内容对某型飞机的所有飞机都有效，则在左下角标有全部（ALL）字样，其示例如图 1-12 所示。

737-600/700/800/900
AIRCRAFT MAINTENANCE MANUAL

LOW RANGE RADIO ALTIMETER (LRRA) SYSTEM - MAINTENANCE PRACTICES

1. **General**

    A. This procedure has one task for the low range radio altimeter (LRRA) system:

       (1) A simulation test of radio altitude.

    TASK 34-33-00-700-801

2. **Radio Altitude Simulation Test**

    A. General

       (1) This task uses an Atlantis DRA707 Radio Altimeter (RA) test set to do a simulation test of radio altitude.

    XXX AIRPLANES WITH PREDICTIVE WINDSHEAR

       (2) You must open the weather radar (WXR) transceiver circuit breaker to make sure that the WXR system does not come on. The radio altimeter supplies radio altitude data to the WXR transceiver. The WXR transceiver uses the radio altitude data to turn the WXR system on and off.

    XXX

    B. References

Reference	Title
24-22-00-860-811	Supply Electrical Power (P/B 201)

    C. Tools/Equipment

    NOTE: When more than one tool part number is listed under the same "Reference" number, the tools shown are alternates to each other within the same airplane series. Tool part numbers that are replaced or non-procurable are preceded by "Opt:", which stands for Optional.

Reference	Description
COM-1922	Test Set - Radio Altimeter (Part #: 110-0430-100, Supplier: 38202, A/P Effectivity: 737-600, -700, -700C, -700QC, -800, -900, -BBJ) (Part #: 110-0460-102-01, Supplier: 38202, A/P Effectivity: 737-300, -400, -500) (Part #: 110-0460-105, Supplier: 38202, A/P Effectivity: 737-300, -400, -500, -600, -700, -700C, -700QC, -800, -900) (Part #: 2041595-5202, Supplier: 41364, A/P Effectivity: 737-600, -700, -700C, -700QC, -800, -900) (Part #: 9599-607-15902, Supplier: F0052, A/P Effectivity: 737-100, -200, -200C, -300, -400, -500, -600, -700, -700C, -700QC, -800, -900) (Part #: DRA707B1, Supplier: 38202, A/P Effectivity: 737-600, -700, -700C, -700QC, -800, -900, -BBJ)

    D. Location Zones

Zone	Area
117	Electrical and Electronics Compartment - Left
118	Electrical and Electronics Compartment - Right
211	Flight Compartment - Left
212	Flight Compartment - Right

    E. Access Panels

Number	Name/Location
117A	Electronic Equipment Access Door

EFFECTIVITY
XXX ALL

34-33-00

Page 201
Feb 10/2006

图 1-12　手册正文页的飞机构型有效性

任务实施

1. 飞机有效性查询

打开前言（Front Matter），单击有效飞机（Effective Aircraft）查询飞机有效性清单，如图 1-13 所示。

AMM 有效性列表中有哪些飞机的身份信息？

737-600/700/800/900 – AMM SHZ D633A101-SHZ Rev 57 - 15 Jun 2015	*BOEING*	Printed by Toolbox: 15 Aug 2015, 08:29:53 PDT DO NOT USE AFTER 14 Sep 2015

EFFECTIVE AIRCRAFT   Issue Date: 15 Jun 2015
EFFECTIVITY: ALL

This manual is applicable to the aircraft in this list:

Model-Series	Operator		Manufacturer			Registration Number
	Identification Code	Effectivity Code	Block Number	Serial Number	Line Number	
737-78S	XXX	001	YA801	30169	631	B-2666
737-78S	XXX	002	YA802	30170	654	B-2667
737-78S	XXX	003	YA803	30171	681	B-2668
737-77L	XXX	009	YA809	32722	1023	B-2669
737-7BX	XXX	706	YA686	30741	823	B-5025
737-7BX	XXX	707	YA687	30742	864	B-5026
737-76N	XXX	720	YA626	29893	710	B-2679
737-79K	XXX	721	YA721	29190	110	B-2633
737-79K	XXX	722	YA722	29191	127	B-2635
737-76N	XXX	727	YA613	32244	895	B-2678
737-87L	XXX	801	YL661	35527	2616	B-5380
737-87L	XXX	802	YL662	35528	2631	B-5381
737-87L	XXX	803	YL663	35529	2677	B-5400
737-87L	XXX	804	YL664	35530	2703	B-5401
737-87L	XXX	805	YL676	35531	2726	B-5402
737-87L	XXX	806	YL677	35532	2851	B-5411
737-87L	XXX	807	YL678	35535	2895	B-5413
737-87L	XXX	808	YL679	35533	2900	B-5412
737-87L	XXX	809	YL680	35534	3003	B-5440
737-87L	XXX	810	YL681	35536	3019	B-5441
737-87L	XXX	811	YL682	39143	3624	B-5606
737-87L	XXX	812	YL683	39144	3643	B-5607
737-87L	XXX	813	YL684	39145	3656	B-5608
737-87L	XXX	814	YL685	39146	3698	B-5612
737-87L	XXX	815	YL686	39147	3705	B-5613
737-87L	XXX	816	YL687	39148	3736	B-5615
737-87L	XXX	817	YL688	39149	3755	B-5616
737-87L	XXX	818	YL689	39150	3770	B-5617
737-87L	XXX	819	YL690	39151	3828	B-5618
737-87L	XXX	820	YQ461	39152	3841	B-5619
737-87L	XXX	821	YS721	39129	4029	B-5670

图 1-13　飞机有效性查询

2. 读懂飞机维护手册飞机有效性列表

机身上蕴藏的国籍秘密

一架飞机从设计、试验、成批生产到投入运行，每一个过程都会有一些代表该过程的编号，用以对机型进行识别，同时也用来确定各类手册、服务通告、适航指令等的有效性。下面以波音飞机维护手册为例来介绍飞机有效性列表中的编号。

● Model Series——机型

● Effectivity Code（客户）——有效性代码

● Identification Code——识别码

● BLOCK NUMBER——批次号

● Serial Number——序列号

● Line Number——生产线号

● Registration Number——注册号

Registration Number 由飞机持有国家的官方指定。开头的一位或两位的字母由国际航空组织给定，代表该航空器注册的国籍，如 N 代表美国，B 代表中国等。飞机的注册号会印在机体上。例如：B-5606（图 1-14），即飞机的注册号，是一架中国的飞机。

图 1-14　B-5606 飞机实物图

因此，飞机 B-5606 的有效性可在图 1-13 中查到，该飞机的有效性代码为 811，批次号为 YL682。

思考题

1. 为什么使用飞机维护手册时需要查询飞机的有效性？

2. 你知道飞机维护手册的飞机有效性列表中能查到哪些重要信息吗？

任务 2 飞机发动机燃油渗漏章节查询

任务描述

在进行飞机维护手册的有效性查询后，飞机发动机燃油渗漏故障维护还需要明确其所在章节。飞机维护手册（AMM）章节按照《ATA 100 规范》编排。机务人员不仅需要熟悉《ATA 100 规范》，还需要熟练掌握 AMM PART I SDS（Systems Description Section）查询确定项目所在章。

《ATA 100 规范》的编排规则是什么？如何通过 SDS 查询确定飞机发动机燃油渗漏故障维护所在章节？

任务要求

- 了解《ATA 100 规范》。
- 了解飞机发动机燃油渗漏故障的 SDS 查询。

知识链接

1.《ATA 100 规范》

美国航空运输协会（Air Transport Association of America，ATA）是一个航空业界的商业协会，它为航空公司制定飞行运作和技术上的标准，是一个跨国航空公司所共同成立的组织。ATA 创立于 1936 年 4 月，由 14 家美国航空公司开会讨论成立，它的宗旨是：帮助航空工业提供世界上最安全的运输；向航空公司提供技术帮助和运营知识以提高它们的安全水平、服务水平和效率；提倡公平的税收和法律环境，培养竞争、健康的航空工业；发展和调节有利于环境、经济合理、技术上可行的工业行为。ATA 得到联邦航空局的支持，发布了很多航空方面的规范，主要涉及电子商务、航空运营、航空安全等规范。ATA 作为航空运行体系的重要构建者，通过自身的不断发展和扩大影响使世界航空工业更加符合航空公司的需要，并不断地提升了航空公司服务水平、技术水平、运行效率、经济效益。

《ATA 100 规范》（Air Transport Association of America Specification No.100）是美国航空运输协会同航空制造商和航空公司共同制定的一种规范，用以统一各种民用航空产品厂商所出版的各种技术资料的编号。《ATA 100 规范》于 1956 年 6 月 1 日首次出版公布，其后进行了数次修订和改版。《ATA 100 规范》将航空器按照系统、结构以及功能进行分类，并配以规定的章节编号，以便于航空器的技术出版物能够统一格式和基本系统分类，使维修人员能够迅速而又准确地查询所需要的技术资料。这一规范已被世界绝大多数国家所接受，它使各国的航空器设计、制造、使用、维护等部门，在各种技术资料、文件、函电和报告等方面统一了编号，从而大大方便了技术交流，促进了航空事业的发展，同时改进了各种资料和文件的归档和管理，为走向国际标准化的资料管理创造了条件。

2.《ATA 100 规范》的章节编码规则

航空设备大体上可分成"航空器"和"动力装置"两大类，其中"航空器"又可划分为"总体""系统"和"结构"三类，"动力装置"则可分为"螺旋桨 / 旋翼"和"发动机"两类，《ATA 100 规范》对每一分类所属各章的编号划分见表 1-1。

表 1-1　《ATA 100 规范》各章编号划分

章区间	类别	内容
05 ～ 12	"总体"类	飞机的总体部分将飞机作为一个总体看待进行工作，分别介绍了飞机时限 / 维护检查、尺寸及区域的划分、顶起支撑、校水平和称重、牵引和滑行、停放标志、铭牌及勤务
20 ～ 49	"系统"类	飞机的系统章节将飞机分成功能不同独立完成相关功能的系统。例如：空调系统、自动驾驶系统、通信系统、导航系统、电源系统、飞行操纵系统、仪表系统等
51 ～ 57	"结构"类	飞机的结构章节包含飞机的舱门、机身、发动机吊舱 / 吊架、安定面、窗户、机翼等
60 ～ 65	"螺旋桨 / 弦翼"类	螺旋桨 / 弦翼章节包含螺旋桨、旋翼、螺旋桨标准施工等
70 ～ 77	"发动机"类	飞机发动机章节包含动力装置、发动机、发动机燃油和控制系统、点火装置、发动机指示系统等

《ATA 100 规范》规定的章节编号的范围从第 5 章到第 91 章。第 1 章到第 4 章是预留给客户的，用于客户编排的维修手册等。《ATA 100 规范》规定的章节编号及其主题名称，见表 1-2。

表 1-2　《ATA 100 规范》的章节编号及其主题名称

章节	主题名称
6	DIMENSION & AREAS 尺寸及区域划分
6-10	Fuselage 机身
6-20	Wing-Stations & Access 机翼站位及接近盖板
6-30	Nacelle/Pylon-Stations & Access 发动机吊舱 / 吊架站位及接近盖板
6-40	Tall Assembly-Stations & Access 尾翼站位及接近盖板
7	LIFTING & SHORING 飞机顶升和支撑
7-10	Jacking 飞机顶升
7-20	Shoring 支撑
8	LEVELING & WEIGHING 校水平和称重
8-10	Weighing & Balancing 载重和平衡
8-20	Leveling 校水平

续表

章节	主题名称
9	TOWING & TAXIING 牵引和滑行
9-10	Towing 牵引
9-20	Taxiing 滑行
10	PARKING & MOORING 停放和系留
10-10	Parking/Storage 停放 / 储存
10-20	Mooring 系留
10-30	Return to Service 返回服务
11	PLACARDS & MARKINGS 铭牌及标志
11-10	Exterior Color Schemes & Markings 外部颜色方案和标志
11-20	Exterior Placards & Marking 外部铭牌及标志
11-30	Interior Placards 内部铭牌
12	SERVICING 勤务
12-10	Replenishing 补充
12-20	Scheduled Servicing 定期勤务
12-30	Unscheduled Servicing 非计划勤务
14	EQUIPMENT OPERATION 设备使用
14-10	Ground Operation Certification 地面使用认证
14-20	Engine Operation/System Check(Runup) 发动机操作 / 系统检查（试车）
14-30	Discrepancy Analysis & Correction(Trouble shooting) 异常分析及纠正（排故）
15	TRAINING OUTLINE 训练大纲
16	GROUND SUPPORT EQUIPMENT 地面支援设备
16-10	Fixed Refuel Equipment 固定加油设备
16-20	Motorized Vehicles 机动车辆
16-30	Towed Vehicles 牵引车辆
16-40	External Ground Power 辅助地面电源
17	FACILITIES EQUIPMENT 设施和设备
17-10	Heating System 加温系统
17-20	Air Conditioner 空调
17-30	Compressed Air 压缩空气
17-40	Air Handling 通风
17-50	Emergency Power 应急电源
17-60	Fire Protection 防火
17-70	Servicing 勤务
17-80	Lifting Equipment 起重设备

续表

章节	主题名称
17-90	Misc. Equipment 其他设备
20	STANDARD PRACTICES-AIRFRAME 施工标准 - 机身
21	AIR CONDITIONING 空调系统
21-10	Compression 压气机
21-20	Distribution 分配
21-30	Pressurization Control 增压控制
21-40	Heating 加温
21-50	Cooling 冷却
21-60	Temperature Control 温度控制
21-70	Moisture/Air Contaminant Control 湿度 / 空气污染控制
22	AUTO FLIGHT 自动飞行
22-10	Autopilot 自动驾驶
22-20	Speed-Attitude Correction(Mach Trim) 速度 - 姿态修正（马赫配平）
22-30	Auto Throttle 自动油门
22-40	System Monitor 系统监控
22-50	Aerodynamic Load Alleviating 减压气动力负荷装置
23	COMMUNICATIONS 通信
23-10	Speech Communications 语音通信
23-15	Satcom 卫星通信
23-20	Data Transmission and Automatic Calling 数据传输与自动呼叫
23-30	Passenger Address & Entertainment 旅客广播和娱乐系统
23-40	Interphone 内话系统
23-50	Audio Integrating 音频综合系统
23-60	Static Discharging 静电放电
23-70	Audio Monitoring(CVR) 音频监听（驾驶舱语音记录器）
23-80	Integrated Automatic Tuning 综合自动调谐
24	ELECTRICAL POWER 电源
24-10	Generator Drive(CSD) 发动机驱动 - 恒速传动装置
24-20	AC Generation 交流发电系统
24-30	DC Generation 直流发电系统
24-40	External Power 外部电源系统
24-50	AC Electrical Load Distribution 交流电源负载分配
24-60	DC Electrical Load Distribution 直流电源负载分配
25	EQUIPMENT/FURNISHINGS 机舱设备 / 内装饰

章节	主题名称
25-10	Flight Compartment 驾驶舱
25-20	Passenger Compartment 客舱
25-30	Buffet and Galley 餐具间和厨房
25-40	Lavatories 盥洗室
25-50	Cargo Compartments 货舱
25-60	Emergency 应急设备
25-70	Accessory Compartments 附件舱
25-80	Insulation 隔热材料
26	FIRE PROTECTION 防火系统
26-10	Detection 探测
26-20	Extinguishing 灭火
26-30	Explosion Suppression 防爆
27	FLIGHT CONTROLS 飞行控制系统
27-10	Aileron and Tab 副翼及调整片操纵
27-20	Rudder/Ruddevator and Tab 方向舵 / 方向舵舵机及调整片操纵
27-30	Elevator and Tab 升降舵及调整片操纵
27-40	Horizontal Stabilizer 水平安定面
27-50	Flaps 襟翼
27-60	Spoiler, Drag Devices and Variable Aerodynamic Fairings 扰流板、减速装置和可变气动整流
27-70	Gust Lock and Damper 突变锁定及阻尼控制
27-80	Lift Augmenting 增升装置
28	FUEL 燃油系统
28-10	Storage 储存
28-20	Distribution 分配
28-30	Dump 泄放
28-40	Indicating 指示
29	HYDRAULIC POWER 液压系统
29-10	Main 主系统
29-20	Auxiliary 辅助系统
29-30	Indicating 指示
30	ICE AND RAIN PROTECTION 防冰和排雨
30-10	Airfoil 翼面
30-20	Air Intakes 进气道

续表

章节	主题名称
30-30	Pitot and Static 动静压
30-40	Windows, Windshields and Doors 窗，风挡和门
30-50	Antennas and Radomes 天线和雷达罩
30-60	Propellers/Rotors 螺旋桨 / 旋翼
30-70	Water Lines 水管
30-80	Detection 探测
31	INDICATING/RECORDING SYSTEMS 指示 / 记录系统
31-10	Instrument and Control Panels 仪表板
31-20	Independent Instruments 独立仪表
31-30	Recorders 记录器
31-40	Central Computers 中央计算机
31-50	Central Warning Systems 中央警告系统
31-60	Central Display Systems 中央显示系统
31-70	Automatic Data Reporting Systems 自动数据报告系统
32	LANDING GEAR 起落架
32-10	Main Gear and Doors 主起落架和舱门
32-20	Nose Gear/Tail Gear and Doors 前 / 尾起落架和舱门
32-30	Extension and Retraction 起落架收放
32-40	Wheel and Brakes 轮组和刹车
32-50	Steering 转弯
32-60	Position ,Warning ,and Ground Safety Switch 位置、警告及接地安全开关
32-70	Supplementary Gear 辅助起落架
33	LIGHTS 灯光
33-10	Flight Compartment and Annunciator Panel 驾驶舱和信号器面板
33-20	Passenger Compartment 客舱灯光
33-30	Cargo and Service Compartments 货舱和勤务舱灯光
33-40	Exterior Lighting 外部照明
33-50	Emergency Lighting 应急照明
34	NAVIGATION 导航
34-10	Flight Environment Date 飞行环境数据
34-20	Attitude and Direction 姿态及指引
34-30	Landing and Taxiing Aids 着陆和滑行辅助
34-40	Independent Position Determining 自主位置判断
34-50	Dependent Position Determining 非自主位置判断

续表

章节	主题名称
46-50	Miscellaneous Information Systems 杂项信息系统
49	AIR BORNE AUXILIARY POWER 机载辅助动力装置
49-10	Power Plant 动力装置
49-20	Engine 发动机
49-30	Engine Fuel and Control 发动机燃油与控制
49-40	Ignition/Starting 点火 / 启动
49-50	Air 空气
49-60	Engine Controls 发动起操纵
49-70	Indicating 指示
49-80	Exhaust 排气
49-90	Oil 滑油
51	STANDARD PRACTICES AND STRUCTURES 标准施工和结构
52	DOORS 门
52-10	Passenger/Grew 旅客 / 机组门
52-20	Emergency Exit 紧急出口
52-30	Cargo 货舱门
52-40	Service 勤务门
52-50	Fixed Interior 内部固定装饰
52-60	Entrance Stairs 登机梯
52-70	Door Warning 门警告
52-80	Landing Gear 起落架舱门
53	FUSELAGE 机身
53-10	Main Frames 主框架
53-20	Auxiliary Structure 辅助结构
53-30	Plate/Skin 板 / 蒙皮
53-40	Attach Fittings 连接接头
53-50	Aerodynamic Fairings 整流罩
54	MACELLES/PYLONS 吊舱 / 吊架
54-10	Main Frame 主框架
54-20	Auxiliary Structures 辅助结构
54-30	Plates/Skin 板 / 蒙皮
54-40	Attach Fittings 接连接头
54-50	Fillets/Fairings 嵌条整流条 / 整流罩
55	STABILIZERS 安定面

续表

章节	主题名称
55-10	Horizontal Stabilizer 水平安定面
55-20	Elevators-Elevon 升降舵 - 升降机
55-30	Vertical Stabilizers 垂直安定面
55-40	Rudder-Ruddevator 方向舵 - 舵机
56	WINDOWS 窗
56-10	Flight Compartment 驾驶舱窗
56-20	Passenger Compartment 客舱窗
56-30	Door 门上窗
56-40	Inspection and Observation 检查与观察
57	WINGS 机翼
57-10	Center 中央机翼
57-20	Outer 外机翼
57-30	Wing Tip 翼尖
57-40	Leading Edge and Leading Edge Devices 前缘与前缘设备
57-50	Trailing Edge and Trailing Edge Devices 后缘与后缘设备
57-60	Ailerons and Elevons 副翼和升降副翼
57-70	Spoilers 扰流板
57-80	As Required 按需
57-90	Wing Folding System 机翼折叠系统
60	STANDARD PRACTICES-PROPELLER/ROTOR 标准施工 - 螺旋桨
61	PROPELLERS/PROPULSORS 螺旋桨 / 推进器
62	MAIN ROTOR(S) 主螺旋桨
63	MAIN ROTOR DRIVE(S) 主螺旋桨驱动
64	TAIL ROTOR 尾桨
65	TAIL ROTOR DRIVE 尾桨驱动
66	ROTOR BLADE AND TAIL PYLON FOLDING 旋翼叶片和尾塔折叠
67	ROTORS FLIGHT CONTROL 旋翼飞行控制
70	STANDARD PRACTICES-ENGINE 标准施工 - 发动机
71	POWER PLANT GENERAL 动力装置概述
71-10	Cowling 风扇包皮
71-20	Mounts 安装架
71-30	Fireseals and Shrouds 防火封严护罩
71-40	Attach Fittings 接连接头
71-50	Electrical Harness 电气导线束

续表

章节	主题名称
71-60	Engine Air Intakes 发动机进气道
71-70	Engine Drains 发动机排放
72	ENGINE(TURBINE/TURBOPROP) 发动机（涡轮 / 涡桨）
72-10	Reduction Gear and Shaft Section 减速机匣
72-20	Air Inlet Section 进气部分
72-30	Compressor Section 压气机部分
72-40	Combustion Section 燃烧室
72-50	Turbine Section 涡轮部分
72-60	Accessory Drives 附件传动
72-70	By-Pass Section 外函道部分
72-80	Propulsor Section (Rear Mounted) 推进器部分（后置）
73	ENGINE-FUEL AND CONTROL 发动机燃油和控制系统
73-10	Distribution 燃油分配
73-20	Controlling 燃油控制
73-30	Indicating 指示
74	IGNITION 点火
74-10	Electrical Power Supply 电源
74-20	Distribution 分配
74-30	Switching 转换
75	BLEED AIR 空气系统
75-10	Engine Anti-Icing 发动机防冰
75-20	Engine Cooling 发动机冷却
75-30	Compressor Control 压气机引气控制
75-40	Indicating 指示
76	ENGINE CONTROL 发动机控制系统
76-10	Power Control 动力控制
76-20	Emergency Shutdown 紧急停车
77	ENGINE INDICATING 发动机显示
77-10	Power 动力
77-20	Temperature 温度
77-30	Analyzers 分析器
77-40	Integrated Engine Instrument Systems 综合发动机仪表系统
78	EXHAUST 排气系统
78-10	Collector/Nozzle 集气 / 喷口

续表

章节	主题名称
78-20	Noise Suppressor 消音器
78-30	Thrust Reverser 反推器
78-40	Supplementary Air 辅助空气
79	OIL 滑油
79-10	Storage 储存
79-20	Distribution 分配
79-30	Indicating 指示
80	STARTING 起动系统
80-10	Cranking 手摇起动
81	TURBINE 涡轮
81-10	Power Recovery 功率恢复
81-20	Turbo-Supercharger 涡轮增压器
82	WATER INJECTION 喷水
82-10	Storage 储存
82-20	Distribution 分配
82-30	Dumping and Purging 排放与清洁
82-40	Indicating 指示
83	ACCESSORY GEAR BOXES 附件齿轮箱
83-10	Drive Shaft Section 传动轴部段
83-20	Gear Box Section 齿轮箱部段
84	PROPULSION AUGMENTATION 推进增强
84-10	JET ASSISTED TAKEOFF 喷气式飞机辅助起飞
91	CHARTS 图表

　　《ATA 100 规范》将每一类的主题内容编入相应的"章"（Chapter），每个主题内容又可以细分为若干个子课题，每一子课题编入相应的"节"（Section），每个子课题又可以再细分为若干个细课目，每一细课目编入相应的"目"（Subject）。这样就将某个主题内容分为大、中、小三个等级并配以对应的章、节、目编号分层次编排。

　　《ATA 100 规范》的章节目编码规则使其内容编排清晰同时方便查找。其章节目的编号由六位数字组成，这 6 位数字分为三个单元，每单元两位数字，以"**-**-**"的形式表示。左起第一单元的数字表示章号，用来表达航空器大的维护、修理项目、飞机系统、机身、机翼、安定面以及发动机等飞机大部件主题内容；第二单元数字表示节号，用来表达子课题内容；第三单元数字表示目号，用来表达细课目的内容，如图 1-15 所示。

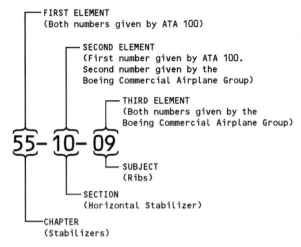

图 1-15　《ATA 100 规范》的章节目编码规则

左起第一单元数字表示章，"55"表示为飞机的安定面；第二单元表示节，"10"表示水平安定面；第三单元表示目，"09"表示为肋。上述表示节号的第二单元数字，其第二位数字可由飞机制造厂家给定并且确定其内容；表示目号的第三单元两位数字由飞机制造厂家给定并确定其内容。这样就使得不同的飞机制造厂家编排的维修手册在大的方面是相同的，但在某些细节方面存在差异，例如，波音飞机的各种维修手册与空客飞机的各种维修手册存在一定的差异。即使同一厂家的飞机，各型飞机之间，甚至同一型号不同改进型之间的手册也有细小的差异。

3. SDS 的功能与结构

SDS（Systems Description Section）是波音系列飞机维护手册的第 I 部分，即 AMM PART I。SDS 是由波音公司遵照《ATA 100 规范》编写的客户化手册。SDS 手册描述了飞机系统的功能、系统的构造、与其他系统的接口、系统部件，以及系统如何操作、维护测试等。国内很多单位将 SDS 手册作为相应机型的培训教材，用于讲解航线或者机库维护工作中日常操作的各设备和各系统的相关知识点。

SDS 的手册结构（图 1-16）和其他手册相同，Front Matter 中包含了本册的有效页清单、修订改版说明、服务通告（SB）等。SDS 手册遵照《ATA 100 规范》划分章节。每个章节代表不同的系统功能，各章中根据系统的功能，又划分为若干子系统、分系统。其章、节、目的编号由六位数字组成，这六位数字分为三个单元，每单元两位数字，以"××-××-××"的形式表示，用户可

```
Legal Notice
Transmittal Letter
Temporary Revisions
Introduction
List of Effective Airplanes
List of Service Bulletins
11 PLACARDS AND MARKINGS
21 Air Conditioning
22 Autoflight
23 Communications
24 Electrical Power
25 Equipment and Furnishings
26 Fire Protection
27 Flight Controls
28 Fuel
29 Hydraulic Power
30 Ice and Rain Protection
31 Indicating and Recording Systems
32 Landing Gear
```

图 1-16　SDS 手册结构

以定义更多功能的子系统。其中，"-00-00"表示该章节下的信息适用于整个系统。例如：如图 1-17 所示，78-00-00（排气）这个章节用来简要描述 78 章中各子章节的内容。

```
□ 78 Engine Exhaust  - CFM56 ENGINES (CFM56-7)
   □ 78-00 Engine Exhaust System
      □ 78-00-00 Engine Exhaust System
         78-00-00-000 ENGINE EXHAUST SYSTEM - INTRODUCTION
         78-00-00-002 ENGINE EXHAUST SYSTEM - GENERAL DESCRIPTION
   ⊞ 78-11 Turbine Exhaust System
   ⊞ 78-30 Thrust Reverser System
   ⊞ 78-31 Thrust Reverser (Introduction)
   ⊞ 78-34 Thrust Reverser Control System
   ⊞ 78-36 Thrust Reverser Indicating System
```

图 1-17　SDS 78 章节结构

以查询 HF 高频通信的功能为例，打开 SDS 手册 23-11-00-001 - HIGH FREQUENCY COMMUNICATION SYSTEM - INTRODUCTION，在 General 中，简单介绍了 HF 通信系统的功能。如图 1-18 所示，HF 通信系统提供了长距离的话音通信，可以实现飞机和飞机之间、飞机和地面之间的通信。此外，还可以看到 HF 通信系统的工作频率、传播方式等。

BOEING　737-600/700/800/900 AIRCRAFT MAINTENANCE MANUAL

HIGH FREQUENCY COMMUNICATION SYSTEM - INTRODUCTION

General

The high frequency (HF) communication system supplies voice communication over long distances. It gives communication between airplanes or between ground stations and airplanes.

The HF system operates in the aeronautical frequency range of 2 MHz to 29.999 MHz. The system uses the surface of the earth and an ionized layer to cause a reflection (skip) of the communication signal. The distance between skips changes due to the time of day, radio frequency, and airplane altitude.

Abbreviations and Acronyms

- ACARS - aircraft communications addressing and reporting system
- ACP - audio control panel
- AM - amplitude modulated
- AME - amplitude modulation equivalent
- ARINC - Aeronautical Radio Incorporated
- BITE - built-in test equipment
- comm - communication
- EE - electronic equipment
- EEC - electronic equipment compartment
- FDR - flight data recorder
- FDRS - flight data recorder system
- freq - frequency
- HF - high frequency
- I/C - interphone communication
- LCD - liquid crystal display
- LED - light emitting diode
- LRU - line replaceable unit
- mic - microphone
- PSEU - proximity switch electronics unit
- PTT - push-to-talk
- REU - remote electronics unit
- RF - radio frequency
- R/T - receive/transmit
- SELCAL - selective calling
- sq - squelch
- sql - squelch
- SSB - single side band
- USB - upper side band
- VSWR - voltage standing wave ratio
- xmit - transmit

EFFECTIVITY
XXX ALL

23-11-00

图 1-18　HF 系统介绍

4. 飞机发动机燃油渗漏故障

发动机漏油可怕吗？

发动机漏油是飞机运行中的常见故障，油液大量渗漏会直接危及飞行安全。在航线维护中，判断发动机漏油以 AMM71-71-00-601 的渗漏标准为依据，如发生渗漏情况，在标准范围内，可以放行飞机：发动机启动好后，在大功率状态下连续渗漏，应该让发动机立即停车进行检查，首先要判断到底是什么类型的油，打开发动机风扇整流罩，由于排放管集中，很难辨明是哪一个地方，这时要眼、鼻并用，找到渗漏源。有时，目视也会判断错误，对发动机进行试车是个好办法，但是浪费时间和燃油，用干净的纸伸到排放管里面，观察里面是否有油迹，这是航线判断发动机漏油常用的方法。

发动机使用的油液通常有液压油、滑油、燃油。液压油渗漏通常最好判断，发动机只有发动机驱动泵和反推系统使用液压油，反推系统的反推作动筒和同步软轴管是最常见的渗漏点。滑油渗漏源主要有滑油/燃油热交换器、前集油槽、后集油槽、启动机、CSD。通过观察发动机渗漏的工作状态对于分析渗漏也很有帮助，如果是启动机漏油，在发动机启动时漏油而启动机脱开后会停止，或者变少。如果是附件齿轮箱的碳封严漏油，漏油会随着发动机的启动越来越严重，因为随着发动机功率增加，发动机带动附件齿轮箱传递的工作载荷增大，齿轮箱渗漏会加剧；如果是 CSD 本身漏油，可以通过观察窗观察油量的减少来判断。燃油渗漏的确定比较复杂。燃油排放管连接着燃油泵、CSD、燃油/滑油热交换器、可调放气活门（VBV）、可调静子导向叶片（VSV）、高压涡轮间隙控制活门（HPTCCV）等众多部件，而且放油管比较集中，要细心观察，不论是任何地方的渗漏，是显性的还是隐藏的，通过试车能直接分辨出渗漏源，如果是少量漏油，漏油不明显，试车解决不了，可以采取相应措施对发动机漏油进行监测。

发动机漏油是很常见的故障，在排除渗漏后，更应该找出渗漏的原因，应该严格执行飞机日常维护工卡制度，部件安装程序要严格按照 AMM 手册来执行，减少发动机的漏油故障的发生，在发生了漏油后，也应该集思广益，采取多种方法判断漏油源，从而找到排除故障的最快方法。

任务实施

1. 飞机发动机燃油渗漏章节查询

以《ATA 100 规范》为依据的飞机维护手册的发动机相关章为 71 章。打开 AMM PART Ⅰ SDS（Systems Description Section）的 71 章目录，确认发动机燃油渗漏所在章节，如图 1-19 所示。

2. 飞机发动机燃油渗漏 SDS 查询

SDS 中可以找到哪些发动机排放的相关信息？

本项目飞机发动机类型为 CFM56-7B，该型发动机油液包含滑油、燃油、液压油等。打开 AMM PART Ⅰ SDS 的 71 章（图 1-20），阅读概述（General）可知，机务维护过程中，发动机的油液排放要避免与发动机发热区域接

触，可通过发动机油液排放来检测组件故障。如果需要查询发动机渗漏允许门限值可查询 AMM PART Ⅱ PP（Practices and Procedures）的 71 章。

BOEING CFM56 ENGINES (CFM56-7) **737-600/700/800/900 AIRCRAFT MAINTENANCE MANUAL**

CHAPTER 71
Power Plant

71-CONTENTS

图 1-19　PP 71 章目录

BOEING CFM56 ENGINES (CFM56-7) **737-600/700/800/900 AIRCRAFT MAINTENANCE MANUAL**

POWER PLANT - ENGINE DRAINS

General

Engine drains prevent fluid contact with hot engine areas. You use engine drains to detect component failures. Engine drains direct these items overboard:

- Oil
- Fuel
- Hydraulic fluid
- Water
- Vapor.

These components drain fluids through the starter air discharge duct in the right fan cowl:

- Strut
- Main oil/fuel heat exchanger
- Hydromechanical unit (HMU)

XXX ALL PRE SB CFM56-7B 73-44
- Burner staging valve (BSV)

XXX ALL
- High pressure turbine active clearance control (HPTACC) valve
- Low pressure turbine active clearance control (LPTACC) valve
- Left and right variable stator vane (VSV) actuators
- Left and right variable bleed valve (VBV) actuators
- Transient bleed valve (TBV).

Fluids drain through a hole in the left fan cowl panel from these components:

- Fuel pump
- Integrated drive generator (IDG)
- Hydraulic pump.

The oil tank drains fluid through a hole in the right fan cowl panel.

See the AMM for more information about allowable leakage limits. (AMM PART II 71-71)

EFFECTIVITY
XXX ALL

71-00-00

Page 12
Jun 10/2003

图 1-20　71 章系统描述

因此，可确认飞机发动机燃油渗漏故障维护程序在 71 章。

思考题

1. 你知道飞机发动机滑油渗漏在哪章吗？
2. 你觉得完成本任务的难点是什么？怎样才能克服困难？
3. 请使用 SDS 查询飞机航行灯的相关信息。

任务 3　飞机发动机燃油渗漏分析

任务描述

通过飞机发动机（CFM56 7B）渗漏故障的 SDS 查询，我们了解到查询发动机渗漏允许门限值需查询 AMM PART Ⅱ PP（Practices and Procedures）的 71 章。由此，为了查询项目故障维护的标准和程序，我们还需要继续探寻飞机发动机燃油渗漏所在节。

PP 手册与 SDS 手册的异同点是什么？如何通过 PP 查询确定飞机发动机燃油渗漏故障维护所在节？

任务要求

- 了解 PP 手册的功能与结构。
- 了解 PP 手册页码段的划分。
- 了解飞机发动机燃油渗漏故障的 PP 查询。

知识链接

1. PP 手册的功能与结构

PP（Practices and Procedures）是波音系列飞机维护手册的第 Ⅱ 部分，即 AMM PART Ⅱ。PP 手册是由波音公司遵照《ATA 100 规范》编写的客户化手册。该手册包含了飞机多种勤务工作、系统的检查、修理、更换等航线和定检工作的详细步骤和技术标准，以及相关工具、耗材、安全注意事项等，是飞机维护工作的主要依据手册。

PP 手册包括前言（Front Matter）和正文章节，手册前言的结构与本项目任务一中介绍的 AMM 前言结构相同，此处不做赘述，此处重点介绍 AMM PART Ⅱ（PP）的正文章节结构。PP 的正文章节按《ATA 100 规范》编写。PP 手册每章包含有效性清单、目录和系统内容，如图 1-21 所示。系统内容由三组数字组成：chapter-section-subject，六位号码可以对应到某航线可更换件（Line Replaceable Unit，LRU）。

（1）章节 / 系统层（XX-00-00）。描述系统功能 / 组件之间的关系，以及系统范围和特点。同时对系统内子系统及与其他系统之间的联系也做说明。通常，这一层级的插图给出这个系统主要部件之间关系的简化框图或通用视角的安装位置，如飞控、起落架等。

（2）节 / 子系统层（XX-X0-00）。详细说明子系统的功能 / 自系统组件、组件与其他系统组件之间的关系，以及每个组件的工作范围。

图 1-21　PP 章节目录

（3）分章节 / 分系统层（XX-XX-00）。安装 / 电路层描述在 XX-XX-X0，用于细分组件中部件的安装 / 电路。

2．PP 手册页码段的划分

PP 手册按照工作的不同内容，将页码分成不同的区段。页码的第一位是功能位，代表该页码段的工作内容和性质，后两位是顺序的页码，表明的是每页的排序，页是 PP 手册的基本单位。PP 手册页码段的划分见表 1-3。

表 1-3　PP 手册页码段的划分

5 ~ 80 章中所表达的主题内容	使用的页码范围	使用的图号范围
描述和操作（Description and Operation (D&O)）	Page 1 ~ 99 以 1 开始，按顺序排列	Figure 1 ~ 99 以图 1 开始，按顺序排列
故障诊断（Trouble Shooting (TS)）	Page 101 ~ 199 以 101 开始，按顺序排列	Figure 101 ~ 199 以图 101 开始，按顺序排列
维修实践（Maintenance Practices (MP)）	Page 201 ~ 299 以 201 开始，按顺序排列	Figure 201 ~ 299 以图 201 开始，按顺序排列
维护勤务（Servicing (SRV)）	Page 301 ~ 399 以 301 开始，按顺序排列	Figure 301 ~ 399 以图 301 开始，按顺序排列
拆卸 / 安装（Removal/Installation (R/I)）	Page 401 ~ 499 以 401 开始，按顺序排列	Figure 401 ~ 499 以图 401 开始，按顺序排列
调整 / 测试（Adjustment/Test (A/T)）	Page 501 ~ 599 以 501 开始，按顺序排列	Figure 501 ~ 599 以图 501 开始，按顺序排列
检查 / 检验（Inspection/Check (I/C)）	Page 601 ~ 699 以 601 开始，按顺序排列	Figure 601 ~ 699 以图 601 开始，按顺序排列
清洁 / 喷漆（Cleaning/Painting (C/P)）	Page 701 ~ 799 以 701 开始，按顺序排列	Figure 701 ~ 799 以图 701 开始，按顺序排列
批准的维修（Approved Repairs (AR)）	Page 801 ~ 899 以 801 开始，按顺序排列	Figure 801 ~ 899 以图 801 开始，按顺序排列

以下对表中前 7 项进行简单介绍。

（1）系统的说明及操作部分（Description and Operation），即手册 001～099 页，提出来单独列为 AMM 手册的 Part I 部分。

（2）飞机系统和组件的故障查找和故障隔离（Trouble Shooting），即手册 101～199 页，另外编写了两本手册分别为故障报告手册（FRM）和故障隔离手册（FIM）。故障报告手册是故障发生时，如何使用故障代码等形式进行报告。故障隔离手册用于对故障的分析、隔离和排除。还增加了放行偏离指南（DDG），以对应最低设备清单（MEL）的内容。因此在不同机型的 AMM 手册中，未熟练使用前，应先熟悉各页码段的内容，以方便查询。

（3）维护施工（Maintenance Practices），即 201～299 页，维护程序由任务和子任务组成。对于某一组件来说，它的维护内容相对简单时，所有内容全部列到 MP 中。或某组件的维护内容不能列到其他页码段时，也列到 MP 页码段内。MP 每个程序都包含工作名称、工作原因、工作准备信息、工作准备、工作程序、收尾工作。

（4）勤务（Servicing），即 301～399 页，AMM 的 12 章是勤务工作，包括如燃油、滑油、液压油、水和轮胎气压的更换和充气，它列出了飞机的计划和非计划的勤务内容。勤务工作出现在有些章节中，它提供了维护动作完成后应出现的结果。例如对减震支柱的充气或充油、控制钢索的润滑，以及饮用水系统的消毒工作。

（5）拆装（Removal/Installation），即 401～499 页，提供系统部件、设备的拆卸与安装程序。

（6）调试 / 测试（Adjustment/Test），即 501～599 页，提供所有程序和参数以评估执行功能操作的系统、子系统、单元、部件或零部件之间的连接件的操作有效性和完整性。测试分为操作测试、功能测试与系统测试。

1）操作测试（Operational Test）仅为了证实某系统或组件可正常工作，这些测试除了使用飞机上安装的设备或装置外，不需要专用工具设备，并且，测试项目和机组执行的测试相差无几。组件操作测试不必满足大修或固定维护周期内要求的特定的规范和容限。

2）功能测试（Functional Test）需要判定某系统或组件所有功能正常并符合系统或组件的最小容许要求和设计规范。这些测试可能要地面支援设备，比操测试要更具体、更详细。这部分内容能够涵盖完成该测试，保证系统的可靠性达到一个可接受水平所需要的所有信息。此测试不用参考额外的文件，一般发生在较短的维修周期内。

3）系统测试（System Test）包含所有的调节规范和容限，以保证系统或组件完成最大功效或设计规范。它应在设备齐全的条件下完成，并能重复进行。一般情况下，在大修时进行系统测试。

（7）检验 / 检查（Inspection/Check），即 601～699 页，确定某个零件、组件、系统的适用性和执行某一功能操作的零件的特殊关系等所需的详细程序。

任务实施

1. 飞机发动机燃油渗漏 PP 查询

打开 AMM PART Ⅱ PP（Practices and Procedures）71 章的目录（Contents），确认该

章包含发动机油液通气与排放（Engine Vents and Drains）相关内容。

如图 1-22 所示，通过目录可确定发动机的通气与排放在 71-71-00，页码为 601。

CFM56 ENGINES (CFM56-7)

BOEING

737-600/700/800/900
AIRCRAFT MAINTENANCE MANUAL

CHAPTER 71
POWER PLANT

71-CONTENTS

图 1-22　71 章系统描述

2. 飞机发动机燃油渗漏检验 / 检查程序查询

如图 1-23 所示，打开 PP 手册 Subject 71-71-00-ENGINE VENTS AND DRAINS，Pageblock 71-71-00-6-Engine Vents and Drains-Inspection/Check，正文显示内容为 CFM56-7 发动机通气与排放的检

发动机燃油渗漏故障检查 / 检验任务单的查询

查 / 检验。值得注意的是，由表 1-3（PP 手册页码段的划分）可知"Pageblock 71-71-00-6"指 71 章 71 节 00 目 601 ～ 699 页码段，对应内容为检验 / 检查（Inspection/Check）工作程序。

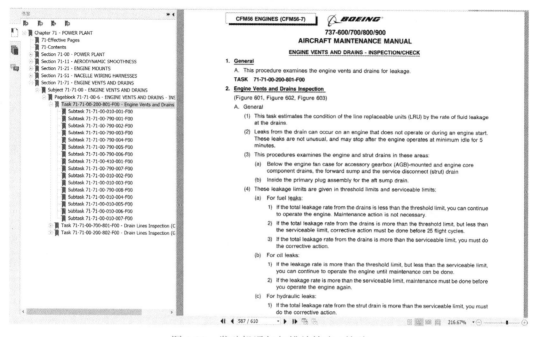

图 1-23　发动机通气与排放检查 / 检验

3. 飞机发动机燃油渗漏分析

PP 手册正文任务卡的阅读是该任务的难点。我们以发动机通气与排放检查任务卡为例来介绍任务卡的结构，以提高工卡阅读的效率。任务卡一级目录分为概述（General）和发动机通气与排放检查

飞机发动机燃油渗漏值在什么范围可以正常放行？

（Engine Vents and Drains）两部分。概述中说明了该任务卡的程序是对发动机的通气与排放渗漏做测试，并给出了任务号。飞机发动机燃油渗漏分析需查询发动机通气与排放检验 / 检查工作程序中的对应子任务。

如图 1-24 所示，由 F 段落中子任务 SUBTASK 71-71-00-790-007-F00 发动机通气与排放渗漏标准与校正动作可知，发动机燃油渗漏系统的放行标准为 180cc/hr（每分钟 60 滴），可服务标准为 270cc/hr（每分钟 90 滴）。如果渗漏小于放行标准，不需要维护；如果放行标准小于渗漏，渗漏小于可服务标准，替换适用部件前还可以飞不超过 25 个循环。因此案例中飞机发动机燃油渗漏是 26 滴 / 分钟，未超过放行标准，不需要维护。

F. Engine Vents and Drains Leakage Limits and Corrective Action

SUBTASK 71-71-00-790-007-F00

(1) Fuel drain system

    (a) Fluid: Fuel

    (b) Threshold limit: 180 cc/hr (60 drops per minute)

        1) If the leakage is less than the threshold limit, no maintenance action is necessary.

        2) If the leakage is more than the threshold limit and less than the serviceable limit, you can continue the engine in service for not more than 25 flight cycles before you replace the applicable component.

    (c) Serviceable limit: 270 cc/hr (90 drops per minute)

图 1-24　飞机发动机燃油渗漏标准

思考题

1. 你知道 PP 手册中页码段（Pageblock）为"-2"代表什么含义吗？
2. 你能写出发动机通气与排放检查的任务号吗？

工卡实操

本项目配套民航机务维修典型案例——AMM 应用工卡 7 个，专业面向包括 AV 和 ME，难度系数设置了Ⅲ（易）、Ⅱ（中）、Ⅰ（难）三个等级，学习者可按需选用。工卡实操过程中，学习者不仅需要关注修理任务本身，同时需有意识地查询手册中相关任务的保障人身与设备安全的警告和注意事项。Warning 内容与保障人员安全有关；Caution 内容与保障产品和设备的安全有关；Note 内容是对修理任务的解释或补充说明。级别最高的等级是 Warning，但无论是哪个等级，工作中，工作者都需要确认已经查询或知晓了相关注意事项和警告，并在工卡上签署确认，才可开始相应任务。本项目配套工卡详情见表 1-4。

民航机务维修人员必读的 Warning、Caution、Note！

表 1-4　项目一配套工卡列表

工卡序号	项目名称	专业方向	难度系数
AMM-01	飞机发动机燃油渗漏分析	ME	Ⅲ
AMM-02	飞机后登机门（左）站位查询	AV/ME	Ⅱ
AMM-03	飞机在强风下牵引	ME	Ⅱ
AMM-04	飞机液压油箱的勤务	AV/ME	Ⅱ
AMM-05	飞机空调系统空气混合腔的安装	AV/ME	Ⅰ
AMM-06	飞机失速警告系统的测试	AV/ME	Ⅰ
AMM-07	飞机客舱座椅的安装	AV/ME	Ⅰ

➡️ 项目评价

	"项目 1　飞机发动机燃油渗漏分析"过程考核评价标准					
课程：　　　　　授课教师：　　　　　专业：						
考核内容	考核标准	评价方式				
知识点（25 分）： 1. 飞机维护手册（AMM）的结构和内容。 2. AMM 有效性查询方法。 3. SDS 手册中飞机发动机燃油渗漏相关信息查询方法。 4. PP 手册中飞机发动机燃油渗漏故障极限值与处理查询方法。	□全对（25 分）。	中国大学 MOOC 平台评分（100%）				
法规知识点（20 分）： 1. 飞机发动机燃油泵拆装施工程序。 2. 飞机发动机燃油渗漏故障处理施工程序。	□制定的施工工卡完全满足施工程序（20 分）。 每错（漏）一项施工程序扣 5 分，扣完为止。	教师评分（100%）				
技能点（20 分）： 1. 飞机发动机燃油泵拆装标准施工。 2. 飞机发动机燃油渗漏故障处理的标准施工。	□飞机发动机燃油渗漏故障处理过程规范，并完成修理任务（20 分）。 每错一步施工步骤，扣 5 分，扣完为止。	小组自评（10%）	小组互评（20%）	教师评分（50%）	企业导师评分（20%）	
工作作风 / 操作安全（5 分）	□正确做好高空作业安全措施，安全步骤不缺失，计 5 分，否则计 0 分。					
团结协作（5 分）	□与所分配成员相处融洽，积极参与、分工协作，顺利完成任务，得 5 分，否则计 0 分。					
节约意识（5 分）	□合理使用耗材，计 5 分，出现浪费计 0 分。					
6S 管理（5 分）	□按 6S 要求整理操作台、实训场地，计 5 分，否则计 0 分。					
安全意识（5 分）	□项目报告中，精准表达对安全意识的体会得 5 分。	教师评分（100%）				
责任使命感（5 分）	□项目报告中，对责任使命感体会深刻得 5 分。					
职业品格（5 分）	□项目报告中，对职业品格体会深刻得 5 分。					

随手笔记

项目 2

IPC 等手册应用：飞机尾白航行灯的安装

Application of IPC and Other Manuals: Installation of White Position Lights on the Aircraft Tail

夜空中最亮的"星"
——飞机航行灯

 项目导读

　　夜晚的时候，我们仰望天空看见一架架飞机从头顶飞过。它们都闪烁着同样的光芒，左侧的翼尖是红光，右侧的翼尖是绿光，尾部的则是白光（图 2-1）。它们用于判明飞行物为飞机，且指示飞机飞行的方向。因此，它们的名字叫航行灯（Navigation Light/Position Light）。航行灯的照射范围是 110 度，所有的航行灯都必须符合这个标准。在飞行中，从驾驶舱看到前方的飞机如果是右侧闪红光，左侧闪绿光，飞行员就马上可以知道有飞机在和自己相向飞行，需要马上做出反应。一般情况下，两架客机的相对速度约为 1800 千米 / 时，从发现到相遇大约只需 10 秒。尽管有地面管制员对飞机的监控以及TCAS 系统的监视，航行灯对飞行员的警示和进一步判断也极其重要。尾部的白光则是告诉周围的飞机前方有飞机。那么当飞机尾白航行灯出现故障，需要及时更换新件时，我们该怎么查询尾白航行灯的件号呢？

图 2-1　夜空中的飞机

 教学目标

　　★ 掌握零部件图解目录手册（IPC）的功能。
　　★ 掌握零部件图解目录手册（IPC）的结构。

★掌握零部件图解目录手册（IPC）的查询方法。
★掌握零部件图解安装方案的制定。

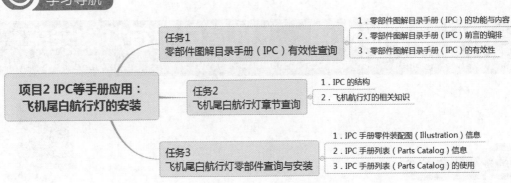

任务 1　零部件图解目录手册（IPC）有效性查询

任务描述

飞机尾白航行灯出现故障，需要通过零部件图解目录查询该飞机尾白航行灯的件号及供应商等信息，及时更换新件，保障飞行安全。零部件图解目录手册（Illustrated Parts Catalog，IPC）是针对飞机上航线可更换部件和组件的购买、装配、储存、生产等编写的一本手册。零部件图解目录手册（IPC）与飞机维修手册（AMM）相似，也是客户化手册，按照《ATA 100 规范》编排。另外，有关电气设备件号的查询除了可利用 IPC 以外，也可利用 WDM，这个知识点在项目 5 中会有详细介绍。

如何查询飞机的有效性呢？ IPC 与 AMM 的飞机有效性查询有何不同呢？

任务要求

● 了解图解零部件目录手册中飞机有效性交互索引。
● 了解图解零部件目录手册中有效性的相关信息。

知识链接

IPC 能干啥？

1. 零部件图解目录手册（IPC）的功能与内容

IPC 手册由零件装配图（Illustration）和零件列表（Parts Catalog）组成。在 IPC 手册中，所有的图和表有对应关系，基本单位是图（Figure），每一个图与相应的零件列表对应，给出部件的件号、名称、部件的供应商、部件的上下级之间的关系、部件之间的互换性等信息。它是客户购买、储存以及更换零部件的依据。

零部件图解目录手册是由飞机生产厂家提供，主要用于航线可更换件的识别、查询、维护及备件的手册。具体来说，IPC 在实际应用中主要有两个功能：①提供可更换零件的位置识别和装配关系等维护信息，为航线维护提供方便；②通过 IPC 手册可以查到与零件相关的数量、有效性、厂家等信息，为航材部门备件计划提供信息。

根据提供手册的厂家不同，常用 IPC 分为三类：

（1）飞机生产厂家提供的飞机图解零件目录（AIPC）。

（2）发动机厂家提供的发动机图解零件目录（EIPC 或 PIPC）。

（3）部件生产厂家提供的部件图解零件目录（IPL）。

2．零部件图解目录手册（IPC）前言的编排

IPC 的前言包含改版传送的信函（Revision Transmittal Letter）、改版传送（Revision Transmittal）、目录（Table of Contents）、介绍（Introduction）、详细零件清单说明（Explanation of Parts List Data）、模块零件清单说明（Explanation of Module Part List Data）、零件定位介绍（Instructions to Locate Part）、图示说明（Explanation of Illustration Techniques）、飞机有效性对照表（Airplane Effectivity Cross Reference）、区域划分图（Zone/Index Diagrams）、段位/站位图（Airplane Section Breakdown/Station Diagrams）、主要图号索引（Major Drawing Numbers Index）、油滤维护工具包清单（Filter Maintenance Kit Listing）、供应商清单（Supplier Name and Address）、服务通告清单（Service Bulletin List）、改装清单（Modification List）、模块交叉索引（Module Cross Reference）、波音规范件对照表（Specification Cross Reference）、件号字母索引表（Part Number Alpha-Numerical Index）、件号数字索引表（Part Number Numerical-Alpha Index）等。IPC 前言编排如图 2-2 所示。

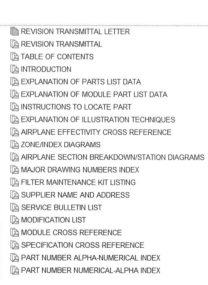

图 2-2　IPC 前言编排

下面对以下几项进行介绍。

（1）飞机有效性对照表（Airplane Effectivity Cross Reference）：显示适用于该手册的所有飞机的各种编号，如图 2-3 所示。

BOEING
737-600/700/800/900
PARTS CATALOG (MAINTENANCE)

XXX   XXX   XXX   AIRLINES

AIRPLANE EFFECTIVITY CROSS REFERENCE

MODEL/ TYPE	CUST EFFECT CODE	CUST EFFECT CODE TERM	VARIABLE ENGR NUMBER	ENGINE SET NUMBER	MFG SERIAL NUMBER	REGISTRY NUMBER	CUST ASSIGNED AIRPLANE IDENT
737-7W0	001		YA811	Y6140	29912	B-2639	
	002		YA812	Y6148	29913	B-2640	
	003		YA813	Y6292	30074	B-2503	
	004	999	YA814	Y6311	30075	B-2502	

图 2-3　飞机有效性对照表

（2）主要图号索引（Major Drawing Numbers Index）：列出飞机上各主要部分和系统的工程图纸号，帮助用户查阅工程图纸，如图 2-4 所示。

BOEING
737-600/700/800/900
PARTS CATALOG (MAINTENANCE)

MAJOR DRAWING NUMBERS INDEX

```
001A0001    AIRPLANE COLLECTOR - MODEL 737-600/700/800/900
001A0101    . FINAL ASSY - PRODUCT COLLECTOR
012A8101    . . FLYAWAY - EQUIPMENT
103A0001    . . SEALING, FINISHES AND DRAINAGE, AIRPLANE
103A0001        INTEGRATION
110A0001    . . WING TO BODY JOIN
113A0001    . . MOVEABLE T.E.
114A0010    . . SECTION 41 FINAL ASSY
115A0005    . . FIXED T.E. TO FINAL ASSY
116A0010    . . FIXED LEADING EDGE - FINAL ASSY
140A0342    . . FUSELAGE FLOOR PANEL
140A0403    . . FUSELAGE AND EMPENNAGE
140A4406    . . FINAL INSTALLATION - SECTION 44
140A4813    . . TAILCONE
141A6050    . . FORWARD ENTRY DOOR - INSTALLATION
141A6100    . . FORWARD ENTRY DOOR - ASSEMBLY
141A6800    . . FORWARD ACCESS DOOR
144A6300    . . EMERGENCY ESCAPE HATCH AND AUTOMATIC
144A6300        OVERWING EXIT
144A6400    . . EMERGENCY ESCAPE HATCH INSTL COMPONENTS
149A7009    . . SECTION 49 SUPPORT STRUCTURE
149A7011    . . SECTION 49 PANEL INSTLS
160A0100    . . FINAL ASSY - LANDING GEAR
170A1601    . . VERTICAL FIN
173A0000    . . RUDDER INTEGRATION
173A0100    . . RUDDER
174A0000    . . DORSAL FIN
180A1601    . . HORIZONTAL STABILIZER
```

图 2-4　主要图号索引

（3）油滤维护工具包清单（Filter Maintenance Kit Listing）：用来协助对飞机油滤的

维护工作。清单中包括了工作包的件号、对应油滤的件号、工作包中的部件和数量等信息，如图 2-5 所示。

```
FILTER MAINTENANCE KIT LISTING


KIT NUMBER
----------
    FOR:                                          KIT CONTENT        QTY
    -----------------------------------------     -----------        ---
63B10463-20
----------------

    RESERVOIR PRESSURIZATION MODULE AIR           NAS1611-012A         1
    FILTER                                        (OPTIONAL P/N:
                                                  NAS1611-012)
    732-11240-01    (SPEC 60B00226-3)             NAS1611-115A         1
    732-11240-02    (SPEC 60B00226-5)             (OPTIONAL P/N·
    732-11240-03    (SPEC 60B00226-7)             NAS1611-115)
    732-11240-04    (SPEC 60B00226-8)             NAS1611-216A         1
    732-11240-05    (SPEC 60B00226-9)             (OPTIONAL P/N:
                                                  NAS1611-216)
```

图 2-5　油滤维护工具包清单

（4）供应商清单（Supplier Name and Address）：给出供应商的名称与地址，如图 2-6 所示。

SUPPLIERS NAME AND ADDRESS
WITH FEDERAL SUPPLY CODE

```
VAA001      SEE V33586

VAB546      IACOBUCCI SPA LOCALITA COLLE BAIOCCO FERENTINO /FR/ 03013
            ITALY

VA4147      UMBRA CUSCINETTI SPA VIA PIAVE 12 06034 FOLIGNO, ITALY

VA5342      AVIOINTERIORS SRL VIA APPIA KM 66400 TOR TRE
            PONTI/LATINA/04013, ITALY FORMERLY VO565B

VA8053      SAIEET SPA VIA DEL VETRAIO 21 BOLOGNA, ITALY 40138

VC0002      NORD-MICRO ELEKTRONIK FEINMECHANIK AG VICTOR-SLOTOSH-STR.
            20 D-6000 FRANKFURT/MAIN 60, FEDERAL REPUBLIC OF GERMANY

VC0029      SEEGER-VERBINDUNGSTECHNIK GMBH ROEMERWEG 118 7033
            HERRENBERG 1, GERMANY

VC0810      HALDER ERWIN MASCHINEN UND WERKZEUGFABRIK POSTFACH 209
            7958 LAUPHEIM 1, GERMANY
```

图 2-6　供应商清单

（5）服务通告清单（Service Bulletin List）与改装清单（Modification List）：列出加入 IPC 中的所有服务通告和改装号及其加入的时间，如图 2-7 所示。

```
                  SERVICE BULLETIN LIST

  SB/L  SERVICE BULLETIN       MOD   MODIFICATION        DATE OF
  FSCM  NUMBER                 FSCM  NUMBER              INCORP

 *A4147 07322-27-01                                      FEB 10/06
  F0559 73-022                                           JAN 10/01
  F1976 75-004                                           JAN 10/02
  F1976 75-025                                           JAN 10/02
  F9588 72-023                                           JAN 10/01
  F9588 72-284                                           JAN 10/02
  S3960 72-065                                           JAN 10/01
  S3960 72-066                                           JAN 10/01
```

图 2-7　服务通告清单与改装清单

（6）波音规范件对照表（Specification Cross Reference）：提供波音规范号和对应供应商件号的相互参照，如图 2-8 所示。

```
             SPECIFICATION NUMBER SEQUENCE

 SPECIFICATION NUMBER    SUPPLIER CODE   SUPPLIER PART NUMBER

 BACP30J10               25337           R30F10
 BACP30J2                22277           BMP30J2
 BACP30J2                25337           R30F2
 BACP30J7                22277           BMP30J7
 BACP30J7                25337           R30F7
 BACP30J8                22277           BMP30J8
 BACP30J8                25337           R30F8
 BACP30J9                22277           BMP30J9
 BACP30J9                25337           R30F9
 BCREF10439              17554           BCREF16530
 BCREF10439              81349           CWR11MH105KB
 BCREF10439              31433           T492B105K035BS
 BCREF10439              2N936           293D035
```

图 2-8　波音规范件对照表

（7）件号字母索引表（Part Number Alpha-Numerical Index）：以字母为序的件号索引表，给出零件所在的章（Chapter）、节（Section）、单元（Unit）、图纸号（Figure）、项目号（Item）、安装个数等信息，如图 2-9 所示。

AACREP4H6FS570

图 2-9 件号字母索引表

PART NUMBER / CH-SECT-UNIT-FIG-ITEM	TTL REQ.	PART NUMBER / CH-SECT-UNIT-FIG-ITEM	TTL REQ.	PART NUMBER / CH-SECT-UNIT-FIG-ITEM	TTL REQ.	PART NUMBER / CH-SECT-UNIT-FIG-ITEM	TTL REQ.	PART NUMBER / CH-SECT-UNIT-FIG-ITEM	TTL REQ.
AACREP4H6FS570		ABR3M3G		ABR4M104		ABW16-101		ACB4-3	
32-44-11 06 65	1	27-41-51 01 45	RF	22-11-26 01 65	RF	52-41-00 01 640	RF	25-24-51 06 110	1
32-44-11 06 120	8	27-41-51 01 45	RF	22-11-26 01 45	RF	52-41-00 01 825	RF	25-24-51 06 110	RF
AACREP4H8FS570		27-41-51 02 45	RF	22-11-26 01 45	RF	52-41-00 05 665	RF	25-24-51 06 110	RF
27-31-34 04 175	2	27-41-51 02 45	RF	22-11-26 01 95	RF	52-41-00 05 840	RF	25-24-51 06 110	RF
AA2100-11D99		52-31-51 02 65	RF	ABR4M104B		ABW4-5		25-24-51 07A 45	1
29-11-61 01 250	4	52-31-51 02 65	RF	22-11-25 01 35	RF	52-13-41 03 45	RF	25-24-51 07A 45	RF
29-11-61 01 750	4	32-51-52 03 165	1	22-11-25 01 65	1	52-13-41 03 100	RF	25-24-51 07A 45	RF
AA4825-121D142		32-51-52 03 165	1	22-11-26 01 45	RF	52-41-41 03 25	RF	25-24-51 07A 45	RF
29-11-61 01 220	1	32-51-52 03 165	RF	22-11-26 01 95	RF	52-41-41 03 70	RF	25-24-51 07A 95	RF
29-11-61 01 720	1	ABR3M5013WGP		ABR4M119		52-41-41 04 55	RF	25-24-51 07A 95	RF
AA55549-13		32-51-52 03 165	RF	27-61-21 02 65	RF	ABW4V5		25-24-51 07A 95	RF
34-43-11 01 20	RF	32-51-52 03 165	1	27-61-21 02 125	RF	52-41-41 05 85	RF	ACB4-3L	
34-43-11 01 20	1	32-51-52 03 165	RF	32-44-11 06 80		52-41-41 05 85	1	25-24-51 07A 95	1
34-43-21 02 20	2	ABR3M5014WGL		ABR4M120		AB4E27			
34-43-21 02 20	RF	32-51-52 03 150	1			49-15-31 01 110	2		
AA6540-121D151						49-15-31 01 120	2		

（8）件号数字索引表（Part Number Numerical-Alpha Index）：以数字为序的件号索引表，如图 2-10 所示。

0-123-000400000

图 2-10 件号数字索引表

PART NUMBER / CH-SECT-UNIT-FIG-ITEM	TTL REQ.	PART NUMBER / CH-SECT-UNIT-FIG-ITEM	TTL REQ.	PART NUMBER / CH-SECT-UNIT-FIG-ITEM	TTL REQ.	PART NUMBER / CH-SECT-UNIT-FIG-ITEM	TTL REQ.	PART NUMBER / CH-SECT-UNIT-FIG-ITEM	TTL REQ.
0-123-000400000		0A254-0033A		0BA1		0L387BPGPL		0P1913-1	
25-31-60 02 272	6	25-11-51 58 296	1	25-24-31 15G 420	RF	33-23-00 63D 96	RF	25-80-00 18 5	80
0-132-001800000		0A254-0034A		25-24-31 15G 420	6	33-23-00 63F 66	62	25-80-00 20 10	RF
25-31-60 02 350	6	25-11-51 58 299	1	25-24-31 15G 425	RF	33-23-00 63F 66	62	25-80-00 20 10	100
0-5-404-604		0A254-0035A		25-24-31 19 215	1	33-23-00 63F 96	62	25-80-00 21 5	2
25-31-60 04 65	204	25-11-51 58 143	1	25-24-31 19 220	RF	33-23-00 63F 96	RF	25-80-00 21 55	RF
0-80		0A254-0036A		25-24-31 19 220	6	0L6832AS15TPL		25-80-00 21 55	7
21-51-02 01 80	4	25-11-51 58 146	1	25-24-31 19 235	RF	33-11-51 05A 45	10	25-80-51 01 5	5
21-51-02 02 80	RF	0A296-0013A		25-24-31 20 274	1	33-11-51 05A 45	RF	25-80-51 01 5	86
21-51-02 02 80	4	25-11-51 58 896	1	25-24-31 20 275	RF	0L7152AS15TPNL		25-80-51 05 5	5
21-51-02 02 80	RF	0A296-0014A		25-24-31 20 275	RF	33-11-51 05A 50	2	25-80-51 05 5	58
0AB20-15		25-11-51 58 898	1	25-24-31 20 280	RF	33-11-51 05A 50	RF	52-21-12 12 35	4
38-31-01 01 20	RF	0A296-0015A		25-24-31 21 210	RF	33-11-51 05A 50	RF	52-21-12 12 35	4
0A011-0236-1		25-11-51 58 780	1	25-24-31 21 210	RF	0MA134		52-21-12 12 315	RF
38-11-01 01 340	1	0A296-0016A		25-24-31 21 210	RF	22-11-37 01 250	RF	52-21-12 12 315	8
0A011-0236-21		25-11-51 58 782	1	25-24-31 21 215	RF	22-11-37 01 250	8	52-21-51 09 7	RF
38-11-01 10 340	1	0A296-0017A							

3. 零部件图解目录手册（IPC）的有效性

每架飞机在自己的机队中都有一个客户有效代码，对应每个代码主要的飞机编号列举在飞机有效性对照表（Airplane Effectivity Cross Reference）里，如图 2-3 所示。

（1）零件目录（Parts Catalog）中每个零件有效性号区段都在有效性栏（EFFECT FROM To）中，如 003099 代表零件适用于 1 号到 99 号飞机（图 2-11）；或多个区段的情况，如 001007/050053 代表零件适用于 1 到 7 号和 50 到 53 号飞机。如果这一栏为空，则表示适用于该手册飞机有效性对照表中提及的所有飞机。

BOEING
737-600/700/800/900
PARTS CATALOG (MAINTENANCE)

FIG ITEM	PART NUMBER	1 2 3 4 5 6 7 NOMENCLATURE	EFFECT FROM TO	UNITS PER ASSY
1		MODULE INSTL-P5 OVHD (CABIN ALTITUDE AND DIFFERENTIAL PRESSURE INDICATOR ONLY)		
- 3	233A3101-40	MODULE INSTL-P5 OVHD (CABIN ALTITUDE AND DIFFERENTIAL PRESSURE INDICATOR ONLY) FOR NHA/OTHER SYS DET SEE: 31-11-00-01B	001002	RF
- 7	233A3101-106	MODULE INSTL-P5 OVHD (CABIN ALTITUDE AND DIFFERENTIAL PRESSURE INDICATOR ONLY) FOR NHA/OTHER SYS DET SEE: 31-11-00-01B	003999	RF
50	AW2835AB06	.INDICATOR-DUAL ALTM AND DIFF PRESSURE SUPPLIER CODE: V61349 FUNCTIONAL DESCRIPTION: DISPLAYS CABIN ALTITUDE TO THE FLIGHT CREW. SPECIFICATION NUMBER: 10-60726-6 ELECTRICAL EQUIP NUMBER: N00051 COMPONENT MAINT MANUAL REF: 21-30-01 MAINTENANCE MANUAL REF: 21-33-02		1
- 60	233A3213-1	.MODULE ASSY-CAB. ALTM AND RATE OF CLIMB P5-16 ELECTRICAL EQUIP NUMBER: P00516 COMPONENT MAINT MANUAL REF: 21-33-21		RF
65	233A3213-101	..MODULE ASSY-		RF

MISSING ITEM NO. NOT APPLICABLE

图 2-11　有效性号区段

（2）IPC 手册中改版信息涉及相关有效性信息。改版信息通过在章节目录表格外左侧加"R"字母标示该行已修改。如图 2-12 所示，章节目录表中除改版信息外，还标识了有效性（EFFECT）号区间。

PARTS CATALOG (MAINTENANCE)

CHAPTER 21 – AIR CONDITIONING

TABLE OF CONTENTS

	SUBJECT	CHAPTER SECTION UNIT	FIGURE	EFFECT
	SERVICE UNIT ASSY-PASS. WITH SPKR AND GASPER (AIR CONDITIONING ONLY)	21-24-00	01	
R	SERVICE UNIT ASSY-PASS. WITH 2K SPKR AND GASPER (AIR CONDITIONING ONLY)	21 24 00	63D	
R	SERVICE UNIT ASSY-PASS. WITH 4K SPKR AND GASPER (AIR CONDITIONING ONLY)	21-24-00	63F	
	STRUCTURE INSTL-LAV A (AIR CONDITIONING ONLY)	21-24-00	07A	
	STRUCTURE INSTL-LAV D AND E (AIR CONDITIONING ONLY)	21-24-00	04	
	STRUCTURE INSTL-LAV E (AIR CONDITIONING ONLY)	21-24-51	04X	
	RECIRCULATION SYSTEM	21-25-00		
	DUCT INSTL-AIR DISTR MIX BAY SYS BASELINE CONFIGURATION (RECIRCULATION FAN CHECK VALVE ONLY)	21-25-03	06	
	DUCT INSTL-AIR DISTR MIX BAY SYS BASELINE CONFIGURATION FOR 737-700 (RECIRCULATION AIR FILTER ONLY)	21-25-01	04	
	DUCT INSTL-AIR DISTR MIX BAY, BASELINE SYS CONFIGURATION FOR 737-700 (RECIRCULATION FAN ONLY)	21-25-02	06	
	EQUIPMENT INSTL-P6 CB PNL (RECIRCULATION FAN ONLY)	21-25-02	02	003999
	EQUIPMENT INSTL-P6 CB PNL (RECIRCULATION FAN ONLY)	21-25-02	25	001002
	JUNCTION BOX ASSY-J24 (RECIRCULATION SYSTEM ONLY)	21-25-00	03	001002
	JUNCTION BOX INSTL-J24 (RECIRCULATION SYSTEM ONLY)	21-25-00	02	003999

图 2-12　目录中的改版信息

任务实施

1. 飞机有效性查询（IPC）

打开前言（Front Matter），单击飞机有效性对照表（Airplane Effectivity Cross Reference）查询飞机有效性代码，如图 2-3 所示。

2. 读懂零部件图解目录手册的飞机有效性对照表

下面以波音飞机维护手册为例介绍零部件图解目录手册的飞机有效性对照表中的编号。

- MODEL/TYPE：机型＋最初拥有这架飞机的用户代码。
- CUST EFFECT CODE：全拼为 Customer Effectivity Code（客户）有效性代码。
- CUST EFFECT CODE TERM：全拼为 Customer Effectivity Code Termination，（客户）有效性终端代码。
- VARIABLE ENGR NUMBER：意为 Variable Engineer Number，批次号。
- ENGINE SET NUMBER：发动机序列号。
- MFG SERIAL NUMBER：全拼为 Manufacturer Serial Number，制造商序列号。
- REGISTRY NUMBER：全拼为 Registration Number，注册号。

以飞机 B-2640 的有效性为例，如图 2-3 所示，该飞机的有效性代码为 002，造册号为 YA812。

思考题

1. 请说说零部件图解目录手册（IPC）与飞机维护手册（AMM）有效性列表的异同点。

2. 你知道零部件图解目录手册的飞机有效性交互索引表中能查到哪些重要信息吗？

任务 2　飞机尾白航行灯章节查询

任务描述

零部件图解目录手册的有效性查询确认了飞机的有效性和适用性。接下来，我们将使用满足飞机有效性的 IPC 手册查询并明确尾白航行灯所在章节。

零部件图解目录手册（IPC）与飞机维护手册（AMM）都属于客户化手册，章节划分按照《ATA 100 规范》编排。我们只要充分熟悉《ATA 100 规范》章节划分的原则，就能快速地查询并确定飞机尾白航行灯所在章。

你还记得《ATA 100 规范》中，飞机上所有的灯光归属的灯光系统在哪章吗？借助已学过的 AMM PART I SDS（Systems Description Section），你是否能获取尾白航行灯的相关信息？

任务要求

- 了解零部件图解目录手册（IPC）的结构。
- 了解飞机尾白航行灯相关信息的查询。

知识链接

1. IPC 的结构

IPC 手册的结构和 AMM 基本相同，Front Matter 中除包含了本册的有效页清单、修订改版说明、服务通告（SB）等，还包括了飞机区域、各站位的划分和图示，以定义部件所在位置。飞机主区域的划分，如图 2-13 所示。

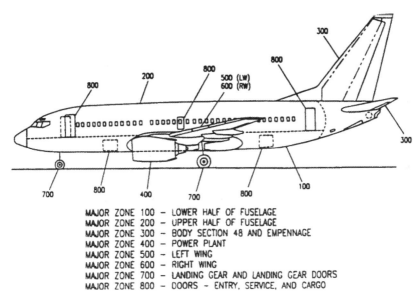

```
MAJOR ZONE 100 — LOWER HALF OF FUSELAGE
MAJOR ZONE 200 — UPPER HALF OF FUSELAGE
MAJOR ZONE 300 — BODY SECTION 48 AND EMPENNAGE
MAJOR ZONE 400 — POWER PLANT
MAJOR ZONE 500 — LEFT WING
MAJOR ZONE 600 — RIGHT WING
MAJOR ZONE 700 — LANDING GEAR AND LANDING GEAR DOORS
MAJOR ZONE 800 — DOORS - ENTRY, SERVICE, AND CARGO
```

FIGURE 1. MAJOR ZONE BREAKDOWN

图 2-13　飞机主区域

IPC 手册与 AMM 类似，按照《ATA 100 规范》编排，由飞机到系统，到组件逐级编写。基本单位为组件，用 6 位数字三个单元"**-**-**"（Chapter-Section-Unit）来表示。不同的是，IPC 手册由图和表的形式呈现。因此，IPC 手册在组件的基础上，将组件分成不同视图，增加了两位数字的图号（Figure），且图中标示的每一个零件使用项目（Item）号，最终在零件列表（Parts Catalog）中逐一对应图中项目号，将具体内容标示到零件，如图 2-14 所示。

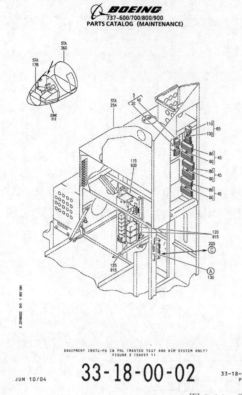

BOEING
737-600/700/800/900
PARTS CATALOG (MAINTENANCE)

FIG ITEM	PART NUMBER	1234567	NOMENCLATURE	EFFECT FROM TO	UNITS PER ASSY
2			EQUIPMENT INSTL-P6 CB PNL (MASTER TEST AND DIM SYSTEM ONLY)	003999	
- 1	233A5220-3		EQUIPMENT INSTL-P6 CB PNL (MASTER TEST AND DIM SYSTEM ONLY) FOR NHA/OTHER SYS DET SEE: 24-30-00-01B	003999	RF
5	69-45109-9		.DIODE ASSY- ELECTRICAL EQUIP NUMBER: M00207 ATTACHING PARTS	003999	1
- 10	WHS10F8TPKL26GY		.SCREW- SUPPLIER CODE: V12324	003999	3
- 15	130069		.NUT- SUPPLIER CODE: V60119 SPECIFICATION NUMBER: BACN10FX16 OPTIONAL PART: RM52LHA4972-6-0 V72962 A11330-6-3 V72962 A4972-6-02 V72962	003999	3
20	1N3311B		..DIODE- SUPPLIER CODE: V81349 ATTACHING PARTS	003999	1
- 25	BACW10P250S		..WASHER	003999	1
- 30	MS35338-44		..WASHER	003999	1
- 35	MS25082-4		..NUT	003999	1
40	69-45109-12		..HEAT SINK	003999	1
45	69-45109-10		.DIODE ASSY- ELECTRICAL EQUIP NUMBER: M00208 M00209 M00210 ATTACHING PARTS	003999	3
- 50	WHS10F8TPKL26GY		.SCREW- SUPPLIER CODE: V12324	003999	12

- ITEM NOT ILLUSTRATED   MISSING ITEM NO. NOT APPLICABLE

EQUIPMENT INSTL-P6 CB PNL (MASTER TEST AND DIM SYSTEM ONLY)
FIGURE 2 (SHEET 1)

JUN 10/04   33-18-00-02   33-18-00-02 PAGE 0   YUN

33-18-00-02

33-18-00 FIG. 2 PAGE 1 JAN 10/03

图 2-14　零部件图解目录手册

　　一般系统的装配顺序遵循详细部件清单（DLP），在详细部件清单里显示了部件和组件与更高一级安装组件的相互关系，按照下列组装顺序，项目名称前的点指出了约定的等级，如图 2-15 所示。

1 2 3 4 5 6 7

安装（Installation）

.安装的详细部件（Detail Parts for Installation）

.组件（Assembly）

.组件的附件（Attaching Parts for Assembly）

..组件的详细部件（Detail Parts for Assembly）

..子组件（Sub-Assembly）

..子组件的附件（Attaching Parts for Sub-Assembly）

...子组件的详细部件（Detail Parts for Sub-Assembly）

...子 - 子组件（Sub-Sub-Assembly）

...子 - 子组件的附件（Attaching Parts for Sub-Sub-Assembly）

....子 - 子组件的详细部件（Detail Parts for Sub-Sub-Assembly）

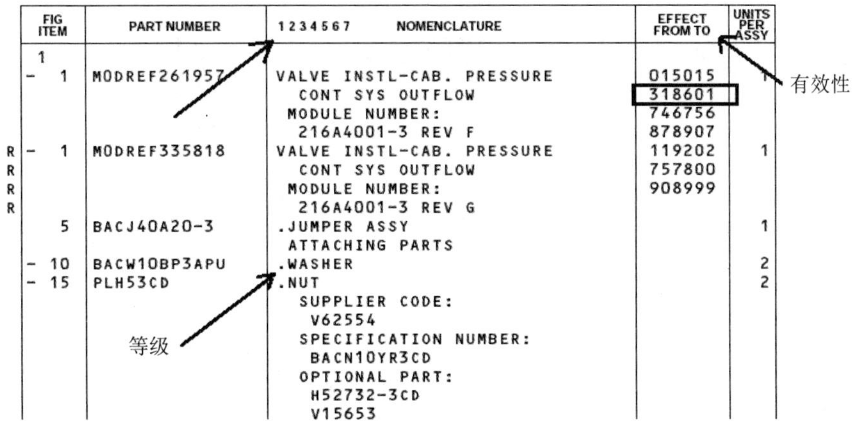

图 2-15 部件等级和有效性

2. 飞机航行灯的相关知识

飞机上为何要装红、绿、白三色灯呢？在世界航空史上，已发生过数百起两机在空中相撞的事故，特别是夜间飞行，如果飞机上没有相应的指示灯更容易发生撞机事件。为了尽量避免这类事故的发生，航空工程技术人员给飞机安装上了红、绿、白三色航行灯。

飞机上的三色航行灯在世界各国有统一的规定：顺着飞机方向看上去，左翼尖端的为红色灯，右翼尖端的为绿色灯，垂尾顶端的为白色灯。为了增大可视程度，有的飞机在翼尖还安装两组同样颜色的灯。飞机驾驶员在驾机飞行的过程中，如果看见前方飞机的航行灯是左红、右绿、中间白时，便知道它是跟自己顺航，也是向前飞行的，只要保持好一定距离便不会相撞；如果看到的是左绿、右红、中间白时，就知道飞机是朝自己迎面飞来的，应立即采取躲避措施，避免两机相撞。

作为飞机照明系统的重要组成部分之一，航行灯工作的可靠性与稳定性保证了飞机在起飞、巡航、着陆等运行阶段的安全性。以派珀 PA44-180 飞机航行灯为例，采用白炽灯为发光源，但是白炽灯内钨丝寿命较短、发热量较高、发光效率较低，并且在高振动环境中经常发生损坏。特别当飞机处于地面低速滑行或停留等待放行阶段时，容易因缺乏足够的冷却气流而导致航行灯灯座发生烧蚀与变形。因此，为保障飞行安全，针对该型飞机开发一套高效率、高安全性、高可靠性的航行灯系统显得尤为重要。

任务实施

1. 飞机尾白航行灯相关信息查询

对于飞机尾白航行灯相关信息，可查询满足飞机有效性的 AMM PART I SDS（Systems Description Section）33 章灯光系统（LIGHTS）- 航行灯（POSITION LIGHTS）- 介绍（INTRODUCTION）。如图 2-16 所示，通过航行灯的显示，可将飞机的位置、方向及姿态信息告知飞机上或地面上的人员。每个翼尖都装有航行灯，它们位于翼尖的前缘或后缘，分别为左红、右绿、尾椎白。航行灯的控制开关在 P5 头顶面板上。

737-600/700/800/900 AIRCRAFT MAINTENANCE MANUAL

LIGHTS - POSITION LIGHTS - INTRODUCTION

Purpose

The position lights show this information to persons in other airplanes or on the ground:

- Airplane position
- Direction
- Attitude.

Physical Description

The position lights are red, green, and white incandescent lights. The left forward position light is red. The right forward light is green. The tail position lights are white.

Location

There are position lights in the tip of each wing. These lights are in the leading and trailing edge of the wing tip.

The control switch for the position lights is on the P5 forward overhead panel.

Operation

You use the position lights switch on the P5 forward overhead panel to control the position light operation.

EFFECTIVITY
XXX ALL

33-43-00

图 2-16　航行灯 SDS 查询

2. 飞机尾白航行灯章节查询

如何确定飞机尾白航行灯所在章?

　　根据《ATA 100 规范》，零部件图解目录手册（IPC）灯光系统章节为 33 章。打开 IPC 手册的 33 章目录，确认飞机尾白航行灯所在章节，如图 2-17 所示。

　　因此，可确认飞机尾白航行灯安装涉及的零部件相关信息在 33 章。

思考题

1. 你知道飞机防撞灯在哪章吗？

2. 你能使用 SDS 查询飞机上有几盏防撞灯，分别安装在哪，它们的颜色分别是什么吗？

48

PARTS CATALOG (MAINTENANCE)

CHAPTER 33 – LIGHTS

TABLE OF CONTENTS

图 2-17 IPC 33 章目录

任务 3 飞机尾白航行灯零部件查询与安装

任务描述

飞机尾白航行灯发生故障后，需要查询尾白航行灯零部件件号或可替换件件号、供应商代码等信息，以便领取或购买相关零部件进行更换。上一任务中，我们已经知晓了尾白航行灯的章节。为了查询尾白航行灯的具体信息，我们还需要深入了解飞机尾白航行灯相关零部件装配图（Illustration）和零部件列表（Parts Catalog）。

如何利用 IPC 手册查询并识读尾白航行灯零部件的件号、名称、部件的供应商、部件的上下级之间的关系、部件之间的互换性等信息？

任务要求

- 了解 IPC 手册中图表信息的含义。
- 制订飞机尾白航行灯换件计划。

知识链接

【大国工匠】胡洋："鲲鹏"
机身数字化装配领军人

1. IPC 手册零件装配图（Illustration）信息

IPC 手册的正文部分按《ATA 100 规范》给出每一个功能章节的图解和零件目录。每一章主体部分按章节目顺序给出飞机各组件的图解和对应的详细零件目录。零部件装配图（Illustration）部分表达了组件的安装位置和各相关件的安装关系。

以液压制动系统（Hydraulic Brake System）32-41-00 为例，打开 03 号图，PAGE0，即为液压制动系统面板组件的零部件装配图 32-41-00-03，如图 2-18 所示。

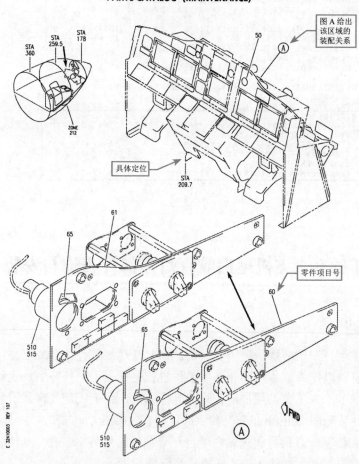

图 2-18　零部件安装图

（1）零部件装配图左上角给出了飞机站位信息图，箭头标注在 STA178 至 STA259.5 之间，零件 A 组件标注了该组件的具体站位信息为 STA209.7。

（2）零部件装配图中，除了给出了组件 A 的具体站位，还给出了零部件的装配关系。

（3）零部件装配图中，所有的零部件标识的是项目号（ITEM），而不是件号（PART NUMBER），例如：图中标识的零部件项目号有 50、60、61、65、510、515 等。

（4）零部件装配图以 0 序号命名，例如：PAGE0、PAGE0A、PAGE0B 等。

（5）零部件装配图与零部件列表（Parts Catalog）相对应，列表按零部件项目号顺序排列，例如，Page1、Page2……等。

2．IPC 手册列表（Parts Catalog）信息

利用 IPC 手册查询并识读飞机尾白航行灯零部件的件号、名称、部件供应商等相关信息，需要掌握 IPC 手册列表（Parts Catalog）中信息的含义、与图的对应关系等。

我们可以借助图 2-19 来对应（Parts Catalog）的相关信息：

（1）图号（FIGURE NUMBER）：图号用于表示飞机某特定部分的图示和零件清单。图号显示在图示标题处，在每页零件清单顶部图号栏内第一个项目号左侧，图号还显示在每个适用页的页脚处。

（2）项目号（ITEM NUMBER）：项目号与图示中零件的项目号相对应。项目号前的横线（一）表示该部件未在图中显示。如"为用户特制目录的说明"中所描述的：部件清单中缺少的项目号是不适用的部件。

（3）高一级组件（NEXT HIGHER ASSEMBLY，NHA）：指由此组件所装配的更大、更高的组件，也可以是分系统。细节（Details）指部件组成所需的更小、更细的零件。Used On：指此组件不能对应所有的 NHA 组件，而是有选择地对应其中某些组件。Altered From：指此组件经过很小的改动变成另一件号代表的组件。

（4）数量信息（UNITS PER ASSEMBLY，UPA）：组件中含有零件的个数，除了可能为数字之外，另有 RF（Reference，参考 NHA 组件中的数量）和 AR（As Required，视装配情况而定）两种内容出现在手册中。

（5）供货商代码（SUPPLIER CODE）：在 NOMENCLATURE（名称）栏内供应商部件号之后标明供应商代码，该信息由 V 字母打头外加五位的数字或者字母指代该供货商的地址、名称信息，可在手册前言部分"供货商的地址、名称信息"中查到。

（6）有效性（EFFECT FROM TO）：表示使用部件的飞机受到限制，而且部件仅适用于特定的一系列飞机。部件的适用性表示飞机可以使用与给出的有效性相对应的部件。因此适用性表示有效性以及可互换性。适用的飞机用 6 位数字代码表示：前三位数字表示可以安装该部件的第一架飞机，后三位数字表示最后一架飞机。例如："742800"表示零件适用于有效性为 742（含）至 800（含）号飞机。当此栏中没有记录时，表示部件适用于 IPC 中所有的飞机。可参照 IPC 中飞机有效性交叉参考部分来确定飞机的有效范围。

FIG ITEM	PART NUMBER	1234567 NOMENCLATURE	EFFECT FROM TO	UNITS PER ASSY
1		VALVE AND WIRE BUNDLE INSTL CAB. PRESSURE CONT SYS OUTFLOW		
R - 1 R	216A4001-1	VALVE INSTL-CAB. PRESSURE CONT SYS OUTFLOW	001012 017019	RF
R R		FOR NHA SEE: 21-30-00-01A	301308 602604	
- 1	MODREF261957	VALVE INSTL-CAB. PRESSURE CONT SYS OUTFLOW MODULE NUMBER: 216A4001-3 REV F	015015 318601 746756 878907	1
R - 1 R R R	MODREF335818	VALVE INSTL-CAB. PRESSURE CONT SYS OUTFLOW MODULE NUMBER: 216A4001-3 REV G	119202 757800 908999	1
5	BACJ40A20-3	.JUMPER ASSY		1
		ATTACHING PARTS		
- 10	BACW10BP3APU	.WASHER		2
- 15	PLH53CD	.NUT SUPPLIER CODE: V62554 SPECIFICATION NUMBER: BACN10YR3CD OPTIONAL PART: H52732-3CD V15653 --------*--------		2
R 20 R R R R R R R R R R	69-41796-1	.SCREW- QUALIFIED I/W DATA: 69-41796-4 WILL NOT FIT 737 CLASSIC. OPTIONAL PART: 69-41796-4	013016 020300 309601 605700 717719 724726 733734 736740 742800 823824 833834 839840	6
55	69-41797-1	.LUG ASSY-		6
60	RM52LHA3022-054	..NUTPLATE- SUPPLIER CODE: V72962 SPECIFICATION NUMBER: BACN10FC5A5 OPTIONAL PART: F1967-5 V15653		1

— ITEM NOT ILLUSTRATED

MISSING ITEM NUMBERS NOT APPLICABLE

21-31-03-01

GEF

21-31-03
FIG. 01
PAGE 2
FEB 15/08

图 2-19　IPC 列表信息

当有设计更改时，将在 NOMENCLATURE 栏内标注有效性的可互换性条件。

IPC 中有可互换性标志的零件之间的关系按以下方式确定：

更换或被更换（REPLACES OR REPL BY）：表示零件（或部件）在物理性和功能性方面的可互换性。

可以使用（MAY USE）：表示根据所含零件向前或向后存在单项可互换性。

与……同 T/W（together with）：表示零件必须与具有互换性的主要部件一同更换，可互换性按照注示完成。

（7）规范编号（SPECIFICATION NUMBER）：该规范号是由波音对组件定义的编号，这个编号详细规定了生产商生产此部件时所有必须具备的性能。这种编号在没有相应供货商取得相关资质时可以代替件号（PART NUMBER）使用，也可以使用 BAC 打头的波音标准件号，因此规范号也可以在件号索引中查到，而且后面还可以跟随取得同样规范的备用件号。

（8）附属部件（ATTACHING PARTS）：紧接着一个项目后列有该项目附属的部件，在部件前面有短语"Attaching Parts"，后面以符号"----*----"结尾。

（9）与其他手册相关部分：WDM 手册中使用的电器设备号 Electrical Equipment Number（后文将会提到）、维修手册与组件相关部分的参考 Maintenance Manual Reference、部件维修手册与组件相关部分的参考 Component Maintenance Manual Reference。

（10）件号信息：超长件号和未设定件号，波音公司的参考文件 Boeing Company Reference（BCREF）为超过 15 位的超长件号和生产厂家未指定件号的设定了件号。BCREF 件号在 PART 栏内记录，而超长件号在 NOMENCLATURE 栏内记录。

（11）图形参考（Illustration reference）：指与该零件功能相关的图示，在所指示图中该零件由虚线画出。

（12）更改标记：更改的数据所在行前面有更改标记，一个字母"R"表示所指的行在当前修改版中被增加或被更改，当一张图的部件清单页包括航空公司的库存号时，而航空公司库存号的更改是该项记录的唯一更改时，所在行前标记有字母"S"。

（13）I/W（Interchangeable With）：指跟随的件号和现有的件号可以互相替换。

3. IPC 手册列表（Parts Catalog）的使用

在维护实践中，使用零部件目录图解手册通常解决两类问题：一类是已知零部件的件号，要确定零部件的名称以及其在飞机系统中的位置；另一类是已知零部件的名称或者零件在飞机系统中的位置，要借助零部件目录图解手册确定零部件的件号。针对这两种情况，零部件目录图解手册提供了相应的查找方法。

（1）已知零部件件号，查询零部件名称、位置等信息的方法：零部件的件号是由大写字母和数字组成的一个号码，由零部件供货厂商或者波音给定，用以指代该零部件的名称，同一件号在同一公司的飞机上，指代唯一的零部件或组件。件号索引分为两部分（图 2-20）：字母数字索引（Alpha-Numerical Index）和数字字母索引（Numerical-Alpha Index）。件号的首位是字母的，查字母数字索引；件号的首位是数字的，用数字字母索引。

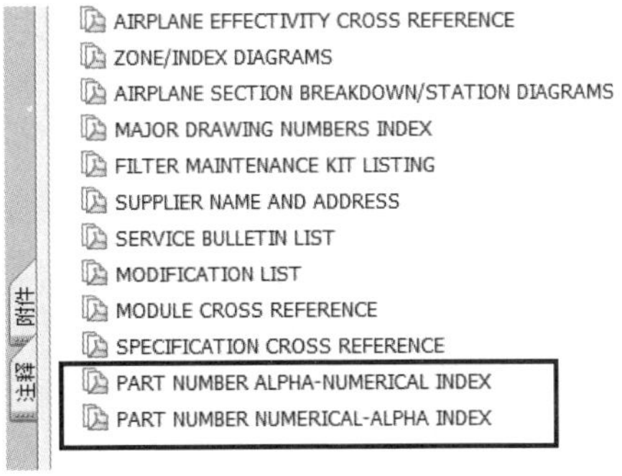

图 2-20　代码索引

1）在 FRONT MATTER 中选择合适的

件号索引（Alpha-Numerical Index/Numerical-Alpha Index）查询待查零部件的件号。

2）记录章（Chapter）- 节（Section）- 单元（Unit）- 图号（Figure）- 项目号（Item）。

3）根据章、节、单元、图号、项目号的信息去查询零部件目录图手册正文。

4）零部件装配图（Illustration）中，由项目号确定待查零部件的安装位置。

5）零部件列表（Parts Catalog）中，由项目号确定待查零部件的件号、名称、供应商代码等相关信息。

（2）已知零部件的安装位置或名称，查询零部件件号的方法：

1）确定章。零部件目录图手册各章节的编号的编排次序完全符合《ATA 100 规范》的规定，可以根据《ATA 100 规范》找到零部件所在的章（Chapter）。

2）确定节。零部件号未知的零件，要借助各章节 Table of Contents 进行查阅，确定零部件可能出现的具体部分（节、组件）。

3）确定图号及组件。各部分的图（Figure）是按字母顺序排列的，找到零部件所在图的标题，记下 Section、Unit、Figure，然后翻到此图，找到零件所在部分的图示，记下 Item 号查询零部件目录列表（Parts Catalog），可知该零件的件号、名称、供应商代码等相关信息。

具体操作的时候，大多数航线可更换件（LRU）的章节号与 AMM 的相应章、节、单元（Chapter-Section-Unit）相同，一般可以利用 AMM 的相应概述部分查找该部件的样貌信息和周围的安装环境，利于针对 IPC 手册图形的检索。因此找到 AMM 的相应部件后即可以到 IPC 手册相应部分寻找，注意不是所有部件都遵循以上规律，附属零件、相关小零件和所安装组件在同一图中查找。

（3）件号未知，且按方法（2）查询未果时，可使用以下查询步骤：当 IPC 手册的编写不规律时，要注意零部件的安装关系和安装位置，这部分内容一般都会在图中标示出来。查找到零件后，一定要到列表中查找相应名称，因为目视是不准确的，经常会把级别不同的零部件，甚至不同的零部件搞混。因此，需要区分机队中，不同飞机对应的图；有时同一架飞机的同一组件的多张图之间，因位置不同或装配分解层次不同，各有区别。另外 IPC 上使用的零部件名称和 AMM 手册中使用的名称不尽相同，因此，要从原理和名称两个方面来进行判断。

1）针对待查零件分析章和系统。

2）查询 AMM 手册的"-00"部分和对应组件部分，注意观察该零件属于什么组件，周围有怎样的连接情况。

3）查询 IPC 手册的相应部分，找到待查零件。

4）如果没有找到，按照连接关系，在相邻组件部分查找。

任务实施

1. 飞机尾白航行灯零部件装配图查询

打开 IPC 33-43-02-01 尾部航行灯组件（LIGHT ASSY-WING TIP AFT POSITION），

单击 PAGE0 查询对应的零部件装配图（Illustration），获得尾白航行灯的站位及灯组 A 装配图。

如图 2-21 所示，通过尾部航行灯装配图可知组件 A 可能为尾白航行灯组件。

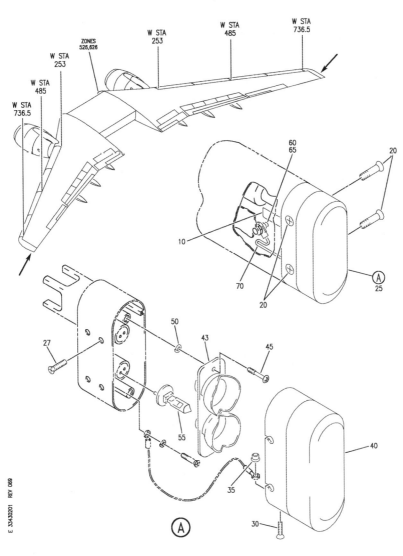

图 2-21　尾部航行灯装配图

2. 飞机尾白航行灯零部件查询与安装

打开 IPC 33-43-02-01 尾部航行灯组件（LIGHT ASSY-WING TIP AFT POSITION）的零部件目录（Parts Catalog），首先需在 PAGE1 中确认零部件装配图的有效性，确认图的可用性后，再对照图和表查询该零部件的具体内容。

如何通过图表确定飞机尾白航行灯的相关信息

如图 2-22 所示，33-43-02 的 1 号图中项目号为 25 的零部件为灯组（LIGHT ASSY-），功能描述（FUNCTIONAL DESCRIPTION）中提到该零部件在飞机尾部闪烁提供白光。因此，可确定该零部件为飞机尾白航行灯。

BOEING
737-600/700/800/900
PARTS CATALOG (MAINTENANCE)

FIG ITEM	PART NUMBER	1234567 NOMENCLATURE	EFFECT FROM TO	UNITS PER ASSY
(1)		LIGHT ASSY-WING TIP AFT POSITION		
- 1	284A1183-1	LIGHT ASSY-WING TIP POSITION POSITION DATA: LH AFT FOR NHA SEE: 57-30-00-01		RF
- 5	284A1183-2	LIGHT ASSY-WING TIP POSITION POSITION DATA: RH AFT FOR NHA SEE: 57-30-00-01		RF
10	BAC27DEX6519	.MARKER- POSITION DATA: LH SIDE PLACARD CONTENT: L00042		1
- 15	BAC27DEX6520	.MARKER- POSITION DATA: RH SIDE PLACARD CONTENT: L00043		1
20	BACB30NN3K3	.BOLT		4
25	30-2481-3	.LIGHT ASSY- SUPPLIER CODE: V72914 FUNCTIONAL DESCRIPTION: PROVIDES WHITE LIGHT SHINING AFT ON TRAILING EDGE OF EACH WING TIP AS PART OF COLOR CODED VISUAL REFERENCE SYSTEM SPECIFICATION NUMBER: S283A313-103 ELECTRICAL EQUIP NUMBER: L00042 L00043 COMPONENT MAINT MANUAL REF: 33-42-06		1
27	60-4991-1	..SCREW-CAPTIVE SUPPLIER CODE: V72914		4
30	MS24693C3	..SCREW		1
35	MS21043-04	..NUT		1
40	31-8300-1	..LENS-WHITE SUPPLIER CODE: V72914		1

MISSING ITEM NO. NOT APPLICABLE

— ITEM NOT ILLUSTRATED

33-43-02-01

33-43-02
FIG. 1
PAGE 1
JAN 10/03

图 2-22　尾部航行灯组件零部件列表

飞机尾白航行灯灯组的件号（PART NUMBER）为 30-2481-3，供应商代码（SUPPLIER CODE）为 V72914，波音标准件号（SPECIFICATON NUMBER）为 S283A313-103，电气设备号（ELECTRICAL EQUIP NUMBER）为 L00042、L00043，组件主要手册参考章节（COMPONENT MAINT MANUAL REF）为 33-42-06，且该灯组在飞机上安装的数量为 1 个。飞机尾白航行灯及灯组具体的安装程序可参考 AMM TASK 33-43-02-960-801 Aft Position Light - Lamp Replacement、AMM TASK 33-43-02-960-802 Aft Position Light - Light Assembly Replacement。

思考题

1. 你能在图 2-21 中找到尾白航行灯灯罩，并说出该零部件的件号、项目号、供应商代码、安装个数等信息吗？

2. 你知道图 2-21 零部件的列表中，项目号为 1、5、15 前面标识的短横线"-"是什么含义吗？

工卡实操

本项目配套民航机务维修典型案例——IPC 应用工卡 6 个，专业面向包括 AV 和 ME，难度系数设置了III（易）、II（中）、I（难）三个等级，学习者可按需选用。工卡实操过程中，学习者不仅需要关注修理任务本身，同时需有意识地查询手册中相关任务的保障人身与设备安全的警告（Warning）和注意事项（Caution/Note）。工作中，工作者需要确认已经查询或知晓了相关注意事项和警告，并在工卡上签署确认，才可开始相应任务。本项目配套工卡详情见表 2-1。

表 2-1　项目 2 配套工卡列表

工卡序号	项目名称	专业方向	难度系数
IPC-01	飞机前厕所烟雾探测组件的安装	AV/ME	III
IPC -02	飞机上 284A2841-1 件的更换	AV/ME	II
IPC -03	飞机上 AS120G120A000 件的更换	ME	II
IPC -04	飞机前轮舱灯组件的更换	AV/ME	I
IPC -05	飞机尾白航行灯的安装	AV/ME	I
IPC -06	飞机垂尾放电刷的更换	AV/ME	I

➡项目评价

考核内容	考核标准	评价方式			
	"项目2 飞机尾白航行灯的安装"过程考核评价标准				
	课程: 授课教师: 专业:				
知识点（25分）： 1. 图解零件目录（IPC）的结构和内容。 2. IPC有效性查询方法。 3. IPC装配图查询方法。 4. IPC中零部件相关信息查询方法。 5. 尾白航行灯相关信息查询方法。	□全对（25分）。	中国大学MOOC平台评分（100%）			
法规知识点（20分）： 1. 尾白航行灯灯泡更换标准施工程序。 2. 尾白航行灯组件更换标准施工程序。 3. 尾白航行灯测试标准施工程序。	□制定的施工工卡完全满足施工程序（20分）。 每错（漏）一项施工程序扣5分，扣完为止。	教师评分（100%）			
技能点（20分）： 1. 尾白航行灯灯泡更换的标准施工。 2. 尾白航行灯组件更换的标准施工。 3. 尾白航行灯测试的标准施工。	□尾白航行灯灯泡更换过程规范，并完成修理任务（20分）。 每错一步施工步骤，扣5分，扣完为止。 没有完成修理任务可按步骤计分，依次为灯泡更换5分，组件更换10分，测试5分。	小组自评（10%）	小组互评（20%）	教师评分（50%）	企业导师评分（20%）
工作作风 操作安全（5分）	□正确做好高空作业安全措施，安全步骤不缺失，计5分，否则计0分。				
团结协作（5分）	□与所分配成员相处融洽，积极参与、分工协作，顺利完成任务，得5分，否则计0分。				
节约意识（5分）	□合理使用耗材，计5分，出现浪费计0分。				
6S管理（5分）	□按6S要求整理操作台、实训场地，计5分，否则计0分				
安全意识（5分）	□项目报告中，精准表达对安全意识的体会得5分。	教师评分（100%）			
责任使命感（5分）	□项目报告中，对责任使命感体会深刻得5分。				
职业品格（5分）	□项目报告中，对职业品格体会深刻得5分。				

项目 3

FIM 等手册应用：飞机防滞 / 自动刹车控制器故障隔离
Application of FIM and Other Manuals: Fault Isolation of Aircraft Anti/Auto Brake Controller

"坚守有我！机务吹响'战疫'号角"——
金鹏航空维修技术骨干严守航空"安全线"

项目导读

　　飞机防滞 / 自动刹车控制器在飞机的起飞、着陆过程中起到至关重要的作用，随着防滞 / 自动刹车控制器的自动化程度越来越高，机务维修岗位的工作职责逐渐转变为保障防滞 / 自动刹车控制器运行的可靠性、稳定性和安全性。这就需要机务维修人员熟悉待修机型防滞 / 自动刹车控制器的控制原理、组成部件，以及常见的故障和排除方法。总体来看，B737-700 型飞机防滞 / 自动刹车控制器可靠性比较高，但传感部分的部件故障偏多，4 个机轮刹车组件使用中可能发生的故障主要有过热、卡滞、漏油、振动等。在飞机再次使用前，机务维修人员如何对其机轮刹车组件进行定期检查？在排除常见的防滞 / 自动刹车控制器故障时，如何利用波音的手册快速、准确地排除故障？

教学目标

　　★掌握故障隔离手册（FIM）的功能。
　　★掌握故障隔离手册（FIM）的结构。
　　★掌握故障隔离手册（FIM）的查询方法。
　　★掌握故障隔离维护方案的制定。

学习导航

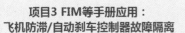

项目3 FIM等手册应用：飞机防滞/自动刹车控制器故障隔离

　任务1　故障隔离手册（FIM）有效性查询
　　1．故障隔离手册（FIM）的功能与内容
　　2．故障隔离手册（FIM）前言的编排

　任务2　飞机防滞/ 自动刹车控制器章节查询
　　1．FIM 的结构
　　2．飞机防滞/ 自动刹车系统的相关知识

　任务3　飞机防滞/ 自动刹车器故障隔离
　　1．FIM 手册维护信息
　　2．故障隔离工卡的阅读

任务1 故障隔离手册（FIM）飞机有效性查询

任务描述

飞机防滞/自动刹车控制器出现故障，需要通过故障隔离手册查询该飞机故障产生可能的原因、相关跳开关、故障隔离程序等信息，以快速、准确地排除故障。故障隔离手册（Fault Isolation Manual，FIM）是针对飞机各系统中需要隔离和排除的故障等编写的一本手册。故障隔离手册（FIM）与飞机维修手册（AMM）和零部件图解目录（IPC）相似，也是客户化手册，按照《ATA 100 规范》编排的。

如何查询飞机的有效性？FIM 与 AMM 和 IPC 的飞机有效性查询有何不同？

任务要求

- 了解故障隔离手册中飞机有效性查询。
- 了解故障隔离手册中有效性的相关信息。

知识链接

FIM 能干啥？

1. 故障隔离手册（FIM）的功能与内容

故障隔离手册（FIM）是由波音商用飞机集团的维护与工程技术服务部排的，与 AMM、IPC 类似，FIM 属于波音公司的客户化手册，手册按照美国航空运输协会《ATA 100 规范》编排。故障隔离手册包含系统中需要隔离和排除的故障，为客户提供一个修理飞机系统或部件故障的方法步骤，以及 737-600/700/800/900 机型设备安装的信息。

FIM 针对飞机有效性清单中列出的飞机所指定用户而制定的内容包括对适用飞机的说明和资料，在用户已交付的构型内加入了适用的波音服务通告或是其他用户对波音通告的一些更改信息，因此在使用时需要注意有效性。

根据故障索引的方式不同，常用 FIM 故障可分为三类：

（1）可观察的故障。可观察的故障通常来自机组报告或维护人员检查，通常伴随着驾驶舱效应，例如驾驶舱故障灯、故障旗、警告信息或者参数错误等。

（2）客舱故障。客舱故障指旅客座舱的故障，如旅客服务组件、舱门、厨房、厕所这些区域设备的失效等。

（3）维护信息。维护信息是通过系统或部件的测试得到的故障信息，有故障代码、故障缩写或故障灯指示。测试可以通过驾驶舱面板进行，也可以在电子设备舱里计算机前面板上进行，或者在驾驶舱，通过控制显示组件 CDU 进行特定系统的测试。

2. 故障隔离手册（FIM）前言的编排

FIM 的前言包含标题页（Title）、飞机有效性对照表（Effective Aircraft）、手册发送说明（Transmittal Letter）、修订重点说明（Highlights）、有效页清单（Effective Pages）、章节有效性（Effective Chapters）、修订记录（Revision Record）、临时

修订记录（Record of Temporary Revisions）、维修服务通告清单（Service Bulletin List）、简介（Introduction）、可观察故障列表（按字母索引）（Observed Fault List: ALPHABETICAL）、可观察故障列表（按系统索引）（Observed Fault List:SYSTEM-ORDER）、客舱故障列表（Cabin Fault List）、客舱故障定位（Cabin Fault Locator）、客舱故障代码索引（Cabin Fault Code Index）等。FIM 前言编排如图 3-1 所示。

FIM Front Matter
- Title
- Effective Aircraft
- Transmittal Letter
- Highlights
- Effective Pages
- Effective Chapters
- Revision Record
- Record of Temporary Revisions
- Service Bulletin List
- Introduction
- Observed Fault List: ALPHABETICAL
- Observed Fault List: SYSTEM-ORDER
- Cabin Fault List
- Cabin Fault Locator
- Cabin Fault Code Index

图 3-1　FIM 前言编排

（1）标题页（Title）：手册的版权页，如图 3-2 所示。

737-600/700/800/900

Fault Isolation Manual

XXX　XXX　XXX　Airlines

BOEING PROPRIETARY, CONFIDENTIAL, AND/OR TRADE SECRET
Copyright © 1997 The Boeing Company
Unpublished Work - All Rights Reserved

Boeing claims copyright in each page of this document only to the extent that the page contains copyrightable subject matter. Boeing also claims copyright in this document as a compilation and/or collective work.

This document includes proprietary information owned by The Boeing Company and/or one or more third parties. Treatment of the document and the information it contains is governed by contract with Boeing. For more information, contact The Boeing Company, P.O. Box 3707, Seattle, Washington 98124.

Boeing, the Boeing signature, the Boeing symbol, 707, 717, 727, 737, 747, 757, 767, 777, BBJ, DC-8, DC-9, DC-10, MD-10, MD-11, MD-80, MD-88, MD-90, and the red-white-and-blue Boeing livery are all trademarks owned by The Boeing Company; and no trademark license is granted in connection with this document unless provided in writing by Boeing.

DOCUMENT D633A103-XXX

ORIGINAL ISSUE DATE: JUNE 5, 1998
PUBLISHED BY BOEING COMMERCIAL AIRPLANES GROUP, SEATTLE, WASHINGTON, USA
A DIVISION OF THE BOEING COMPANY
PAGE DATE: Feb 10/2006

图 3-2　标题页

（2）飞机有效性对照表（Effective Aircraft）：显示适用于该手册的所有飞机的各种编号，如图 3-3 所示。

BOEING®

737-600/700/800/900

FAULT ISOLATION MANUAL

This manual is applicable to the aircraft in this list:

| Model-Series | Operator | | Manufacturer | | | Registration Number |
	Identification Code	Effectivity Code	Block Number	Serial Number	Line Number	
737-7W0	XXX	001	YA811	29912	140	B-2639
737-7W0	XXX	002	YA812	29913	148	B-2640
737-7W0	XXX	003	YA813	30074	292	B-2503
737-7W0	XXX	004	YA814	30075	311	B-2502

EFFECTIVE AIRCRAFT

图 3-3 飞机有效性对照表

（3）手册发送说明（Transmittal Letter）：致手册持有人士的信函，如图 3-4 所示。

737-600/700/800/900
FAULT ISOLATION MANUAL

XXX   XXX XXX   Airlines
XXX
Revision No. 29
Feb 10/2006

To: All holders of this Boeing Document D633A103-**XXX**

Attached is the current revision to the Boeing 737-600/700/800/900 Fault Isolation Manual.

The Fault Isolation Manual (FIM) is furnished either as a printed manual, on microfilm, or digital products, or any combination of the three. This revision replaces all previous microfilm cartridges or digital products. All microfilm and digital products are reissued with all obsolete data deleted and all updated pages added

For printed manuals, changes are indicated on the List of Effective Pages (LEP). The pages which are revised will be identified on the LEP by an R (Revised), A (Added), O (Overflow, i.e. changes to the document structure and/or page layout), or D (Deleted). The pages that contain customer originated data will be identified on the LEP by a C (COC). Each page in the LEP is identified by Chapter-Section-Subject number, page number and page date.

Pages replaced or made obsolete by this revision should be removed and destroyed.

ATTENTION

IF YOU RECEIVE PRINTED REVISIONS, PLEASE VERIFY THAT YOU HAVE RECEIVED AND FILED THE PREVIOUS REVISION. BOEING MUST BE NOTIFIED WITHIN 30 DAYS IF YOU HAVE NOT RECEIVED THE PREVIOUS REVISION. REQUESTS FOR REVISIONS OTHER THAN THE PREVIOUS REVISION WILL REQUIRE A COMPLETE MANUAL REPRINT SUBJECT TO REPRINT CHARGES SHOWN IN THE DATA AND SERVICES CATALOG.

TRANSMITTAL LETTER

图 3-4　手册发送说明

（4）修订重点说明（Highlights）：列出了手册换版时各章节的更改位置和更改原因，如图 3-5 所示。

 BOEING®
737-600/700/800/900
FAULT ISOLATION MANUAL

Location of Change	Description of Change
FRONTMATTER	Changed the List of Service Bulletins.
	Changed reference from ''05-51 TASK 805'' to ''05-51 TASK 802''
	Changed reference from ''34-58 TASK 801'' to ''34-58 TASK 802''
	Changed reference from ''34-58 TASK 802'' to ''34-58 TASK 801''
	Changed reference from ''34-21 TASK 801'' to ''34-21 TASK 835''
CHAPTER 05	
05-51 TASKS	
TASK 809	Changed reference from ''Landing Gear Down Overspeed Condition, Conditional Inspection, AMM TASK 05-51-47-210-801'' to ''Landing Gear Operation Above Design Speed Condition, Conditional Inspection, AMM TASK 05-51-47-210-801''
CHAPTER 21	
21-00 TASKS	
TASK 802	Changed FIM 21-00 Task 802 to reflect that one of the possible causes of the fault condition described would be the right pack, not the left pack.
21-31 TASKS	
TASK 801	Changed Task 801, Table 201 to add and revise data for the BCM fault codes and fault descriptions to reflect the vendor source data.
21-61 TASKS	
TASK 843	Changed the data for the reference.
	Changed reference from ''Cabin Temperature Indicator Removal, AMM TASK 21-61-32-000-801'' to ''Cabin Temperature Bulb Removal, AMM TASK 21-61-31-000-801''
	Changed reference from ''Cabin Temperature Indicator Removal, AMM TASK 21-61-32-000-801'' to ''Cabin Temperature Bulb Installation, AMM TASK 21-61-31-400-801''
TASK 862	Added a possible cause to FIM 21-61, TASK -862 of conditioned air sidewall riser ducts separating from the distribution manifold near the passenger floor level.
CHAPTER 22	
22-11 TASKS	
TASK 801	Added missing maintenance messages CH 22.
TASK 826	Added 22-11-00-810-826, Master FCC cannot enter BITE.
TASK 827	Added 22-11-00-810-827, ADIRU Data Invalid.
22-12 TASKS	
TASK 812	Added steps for checking the Control Column Switching Modules.
TASK 815	Changed the pin call out for 22-12-TASK 815.
22-13 TASKS	
TASK 810	Changed steps in TOGA Switch-1 Problem and TOGA Switch-2 Problem - Fault Isolation.

HIGHLIGHTS

图 3-5　修订重点说明

（5）有效页清单（Effective Pages）：查验章节页是否现行有效，如图 3-6 所示。

BOEING®

737-600/700/800/900

FAULT ISOLATION MANUAL

Page	Date	COC	Page	Date	COC	Page	Date	COC
TITLE PAGE			INTRODUCTION			OBSERVED FAULT LIST - ALPHA (cont)		
O 1	Feb 10/2006		1	Feb 10/2005		R 23	Feb 10/2006	
2	BLANK		2	Feb 10/2005		R 24	Feb 10/2006	
EFFECTIVE AIRCRAFT			3	Jun 10/2005		25	Oct 10/2005	
1	Feb 10/2005		4	Jun 10/2005		R 26	Feb 10/2006	
2	BLANK		5	Feb 10/2005		27	Oct 10/2005	
TRANSMITTAL LETTER			6	Feb 10/2005		28	Oct 10/2005	
O 1	Feb 10/2006		7	Feb 10/2005		29	Oct 10/2005	
2	Feb 10/2005		8	Feb 10/2005		30	Oct 10/2005	
HIGHLIGHTS			9	Feb 10/2005		31	Oct 10/2005	
O 1	Feb 10/2006		10	Feb 10/2005		32	Oct 10/2005	
O 2	Feb 10/2006		11	Feb 10/2005		33	Oct 10/2005	
O 3	Feb 10/2006		12	Feb 10/2005		34	Oct 10/2005	
O 4	Feb 10/2006		13	Feb 10/2005		35	Oct 10/2005	
O 5	Feb 10/2006		14	Feb 10/2005		36	Oct 10/2005	
O 6	Feb 10/2006		15	Feb 10/2005		37	Oct 10/2005	
O 7	Feb 10/2006		16	Feb 10/2005		R 38	Feb 10/2006	
O 8	BLANK		OBSERVED FAULT LIST - ALPHA			39	Oct 10/2005	
D 9	Feb 10/2006		R 1	Feb 10/2006		40	Oct 10/2005	
D 10	Feb 10/2006		2	Oct 10/2005		R 41	Feb 10/2006	
D 11	Feb 10/2006		3	Oct 10/2005		O 42	Feb 10/2006	
D 12	Feb 10/2006		4	Oct 10/2005		O 43	Feb 10/2006	
D 13	Feb 10/2006		5	Oct 10/2005		O 44	Feb 10/2006	
D 14	BLANK		6	Oct 10/2005		O 45	Feb 10/2006	
EFFECTIVE PAGES			7	Oct 10/2005		O 46	Feb 10/2006	
1 thru 3	Feb 10/2006		8	Oct 10/2005		R 47	Feb 10/2006	
4	BLANK		9	Oct 10/2005		O 48	Feb 10/2006	
EFFECTIVE CHAPTERS			10	Oct 10/2005		O 49	Feb 10/2006	
O 1	Feb 10/2006		11	Oct 10/2005		O 50	Feb 10/2006	
2	BLANK		R 12	Feb 10/2006		O 51	Feb 10/2006	
REVISION RECORD			13	Oct 10/2005		O 52	Feb 10/2006	
1	Feb 10/2005		14	Oct 10/2005		O 53	Feb 10/2006	
2	Feb 10/2005		15	Oct 10/2005		R 54	Feb 10/2006	
RECORD OF TEMPORARY REVISIONS			16	Oct 10/2005		O 55	Feb 10/2006	
1	Feb 10/2005		R 17	Feb 10/2006		O 56	Feb 10/2006	
2	Feb 10/2005		O 18	Feb 10/2006		O 57	Feb 10/2006	
SERVICE BULLETIN LIST			O 19	Feb 10/2006		O 58	Feb 10/2006	
O 1	Feb 10/2006		O 20	Feb 10/2006		O 59	Feb 10/2006	
2	BLANK		O 21	Feb 10/2006		O 60	Feb 10/2006	
			R 22	Feb 10/2006		O 61	Feb 10/2006	

A = Added, R = Revised, D = Deleted, O = Overflow, C = Customer Originated

EFFECTIVE PAGES

图 3-6　有效页清单

（6）章节有效性（Effective Chapters）：描述了各章节增加、改动的信息，如图 3-7 所示。

BOEING

737-600/700/800/900

FAULT ISOLATION MANUAL

	Chapter	Date	Title
R	05	Feb 10/2006	TIME LIMITS/MAINTENANCE CHECKS
R	21	Feb 10/2006	AIR CONDITIONING
R	22	Feb 10/2006	AUTOFLIGHT
R	23	Feb 10/2006	COMMUNICATIONS
	24	Oct 10/2005	ELECTRICAL POWER
R	25	Feb 10/2006	EQUIPMENT/FURNISHINGS
R	26	Feb 10/2006	FIRE PROTECTION
R	27	Feb 10/2006	FLIGHT CONTROLS
R	28	Feb 10/2006	FUEL
R	29	Feb 10/2006	HYDRAULIC POWER
R	30	Feb 10/2006	ICE AND RAIN PROTECTION
R	31	Feb 10/2006	INDICATING/RECORDING SYSTEMS
R	32	Feb 10/2006	LANDING GEAR
	33	Oct 10/2005	LIGHTS
R	34	Feb 10/2006	NAVIGATION
	35	Oct 10/2005	OXYGEN
	36	Oct 10/2005	PNEUMATIC
	38	Oct 10/2005	WATER/WASTE
R	49	Feb 10/2006	AUXILIARY POWER UNIT
R	52	Feb 10/2006	DOORS
	56	Oct 10/2005	WINDOWS
CFM56 ENGINES (CFM56-7)			
R	71	Feb 10/2006	POWER PLANT
R	73	Feb 10/2006	ENGINE FUEL AND CONTROL
	74	Oct 10/2005	IGNITION
	75	Oct 10/2005	AIR
R	76	Feb 10/2006	ENGINE CONTROLS
	77	Oct 10/2005	ENGINE INDICATION
R	78	Feb 10/2006	EXHAUST
	79	Oct 10/2005	OIL
R	80	Feb 10/2006	STARTING

A = Added, R = Revised

EFFECTIVE CHAPTERS

图 3-7　章节有效性

（7）修订记录（Revision Record）：用来记录每一次定期修订。飞机维修手册有正常修订和临时修订两种，内容详见项目 1 中介绍。图 3-8 所示为定期修订记录，即为正常修订。

BOEING®
737-600/700/800/900
FAULT ISOLATION MANUAL

All revisions to this manual will be accompanied by transmittal sheet bearing the revision number. Enter the revision number in numerical order, together with the revision date, the date filed and the initials of the person filing.

Revision		Filed		Revision		Filed	
Number	Date	Date	Initials	Number	Date	Date	Initials

REVISION RECORD

图 3-8　定期修订记录

（8）临时修订记录（Record of Temporary Revisions）：用来记录每一次临时修订。在手册两次连续正式修订期间，如果要对手册内容进行修订则进行临时修订。每次临时修订只能修订一个项目的内容，修订后也应做好相应记录，如图 3-9 所示。

BOEING
737-600/700/800/900
FAULT ISOLATION MANUAL

All temporary revisions to this manual will be accompanied by a cover sheet bearing the temporary revision number. Enter the temporary revision number in numerical order, together with the temporary revision date, the date the temporary revision is inserted and the initials of the person filing.
When the temporary revision is incorporated or cancelled, and the pages are removed, enter the date the pages are removed and the initials of the person who removed the temporary revision.

Temporary Revision		Inserted		Removed		Temporary Revision		Inserted		Removed	
Number	Date	Date	Initials	Date	Initials	Date	Initials	Number	Date	Date	Initials

RECORD OF TEMPORARY REVISIONS

图 3-9　临时修订记录

（9）维修服务通告清单（Service Bulletin List）：包括服务通告号、服务通告涉及的 ATA 章、通告状态、出版日期及其对手册有效性的影响等，如图 3-10 所示。

BOEING®
737-600/700/800/900
FAULT ISOLATION MANUAL

Number	Incorporated	Started/Completed	ATA	Subject
A SB 25-1524	OCT 10/2004	S	CHAPTER 52	EQUIPMENT/FURNISHINGS - Flight Compartment Door - Flight Compartment Door Replacement

A = Added, R = Revised

SERVICE BULLETIN LIST

图 3-10　维修服务通告清单

在 Started/Completed 项目中，如果标有 S，则说明这份服务通告尚未完成，如果标有 C，则说明这份服务通告已经完成。

（10）简介（Introduction）：描述了手册包含隔离和纠正安装在 737-600/700/800/900 系列飞机上的系统和设备故障所需的信息，如图 3-11 所示。

BOEING®

737-600/700/800/900

FAULT ISOLATION MANUAL

INTRODUCTION

1. GENERAL

    A. This publication was prepared by the Boeing Commercial Airplane Group in accordance with Air Transport Association of America Specification No. 100, Specification for Manufacturers' Technical Data. It contains information required to isolate and correct faults in the systems and equipment installed in the 737-600/700/800/900 family of airplanes.

    NOTE: "THIS MANUAL IS PREPARED SPECIFICALLY TO COVER THE BOEING AIRPLANES LISTED IN THE 'LIST OF EFFECTIVE AIRPLANES' SECTION, FOR THE OPERATOR NAMED ON THE TITLE PAGE. IT CONTAINS INSTRUCTIONS AND INFORMATION APPLICABLE SOLELY TO THOSE SPECIFIC AIRPLANES, IN THEIR AS-DELIVERED CONFIGURATION PLUS ANY APPLICABLE BOEING SERVICE BULLETINS OR OTHER OPERATOR CHANGES THE INCORPORATION OF WHICH THE NAMED OPERATOR HAS NOTIFIED BOEING."

    NOTE: "THE NAMED OPERATOR IS SOLELY RESPONSIBLE FOR THE ACCURACY AND VALIDITY OF ALL INFORMATION FURNISHED BY THAT NAMED OPERATOR OR ANY OTHER PARTY BESIDES BOEING AND, IF IN RECEIPT OF ACTIVE REVISION SERVICE, THAT ANY MODIFICATIONS TO THE AIRPLANES ARE PROPERLY REFLECTED IN THE MAINTENANCE INSTRUCTIONS CONTAINED IN THIS MANUAL. OPERATORS ARE RESPONSIBLE FOR ENSURING THAT THE MAINTENANCE DOCUMENTATION THEY ARE USING IS COMPLETE AND MATCHES THE CURRENT CONFIGURATION OF THE AIRPLANE. THE BOEING COMPANY ASSUMES NO RESPONSIBILITY IN THIS REGARD. CUSTOMIZATION DOES NOT TRACK THE CONFIGURATION OF AIRCRAFT LISTED ON THE LIST OF EFFECTIVE AIRPLANES PAGE THAT HAVE BEEN CONVEYED TO ANOTHER OPERATOR."

    NOTE: "THIS MANUAL IS NOT SUITABLE FOR USE, INCLUDING WITHOUT LIMITATION, GENERAL INSTRUCTIONS OR TRAINING, FOR ANY AIRPLANES NOT LISTED HEREIN, NOR DOES IT NECESSARILY APPLY TO LISTED AIRPLANES THAT HAVE BEEN CONVEYED TO OTHER OPERATORS."

INTRODUCTION

图 3-11　简介

（11）可观察故障列表（按字母索引）（Observed Fault List: ALPHABETICAL）：所有可观察故障按字母顺序列表，为每一个故障提供相对应的故障代码和故障隔离手册的工卡号，如图 3-12 所示。

BOEING®
737-600/700/800/900
FAULT ISOLATION MANUAL

FAULT DESCRIPTION	FAULT CODE	GO TO FIM TASK
A/P amber warning annunciator light		
• light on flashing, autopilot went to CWS mode	221 020 00	22-11 TASK 801
A/P red warning annunciator light		
• light on steady.	221 010 00	22-11 TASK 801
A/T amber warning annunciator light		
• light on flashing.	223 120 00	22-32 TASK 801
A/T LIMIT indication shows on autothrottle thrust mode display .	223 220 00	22-32 TASK 815
AC and DC metering panel		
• has missing segments	243 030 00	24-31 TASK 827
ACARS		
• does not operate correctly	232 310 00	23-27 TASK 801
Accessory compartment light		
• does not come on	333 050 00	33-30 TASK 801
• does not go off	333 051 00	33-30 TASK 801
ADF		
• audio problem		
- no. 1.	345 310 01	34-57 TASK 806
- no. 2.	345 310 02	34-57 TASK 807
• bearing pointer 1 does not show, ADF flag does not show		
- no. 1.	345 370 01	34-57 TASK 820
• bearing pointer 2 does not show, ADF flag does not show		
- no. 2.	345 370 02	34-57 TASK 821
• indication blank		
- no. 1 captain's	345 320 31	34-57 TASK 808
- no. 1 captain's and first officer's	345 320 48	34-57 TASK 810
- no. 1 first officer's	345 320 32	34-57 TASK 809
- no. 2 captain's	345 330 31	34-57 TASK 811
- no. 2 captain's and first officer's	345 330 48	34-57 TASK 813
- no. 2 first officer's	345 330 32	34-57 TASK 812
• tone problem		
- no. 1.	345 340 01	34-57 TASK 814
- no. 2.	345 340 02	34-57 TASK 815

┌─ EFFECTIVITY ─────────────
│ **XXX ALL**

OBSERVED FAULT LIST
ALPHABETICAL

图 3-12　可观察故障列表：按字母索引

（12）可观察故障列表（按系统索引）（Observed Fault List: SYSTEM-ORDER）：所有可观察故障列入适用的飞机系统中，按《ATA 100 规范》系统章节顺序列表，如图 3-13 所示。

BOEING®
737-600/700/800/900
FAULT ISOLATION MANUAL

FAULT DESCRIPTION	FAULT CODE	GO TO FIM TASK
05-5 TIME LIMITS/MAINTENANCE CHECKS - UNSCHEDULED MAINTENANCE CHECKS		
Airframe		
• vibration or lateral oscillation is excessive	055 170 00	05-51 TASK 813
Birdstrike/FOD		
• on airframe/area unknown, engine parameters normal . .	055 130 00	05-51 TASK 804
Brakes		
• overheated during stop	055 030 00	05-51 TASK 805
Dust (extreme) condition	055 020 00	05-51 TASK 802
Engine nacelle or strut dragged or damaged		
- engine 1	055 100 51	05-51 TASK 811
- engine 2	055 100 52	05-51 TASK 811
Engine seizure		
- engine 1	055 120 51	05-51 TASK 811
- engine 2	055 120 52	05-51 TASK 811
Hard landing or high drag/side load landing	055 040 00	05-51 TASK 806
Lightning strike	055 010 00	05-51 TASK 801
Overspeed		
• with flaps/slats extended	055 090 00	05-51 TASK 810
• with landing gear down	055 080 00	05-51 TASK 809
Overweight landing	055 060 00	05-51 TASK 807
Tail dragged	055 110 00	05-51 TASK 812
Turbulence (severe), stall, buffet, or overspeed	055 070 00	05-51 TASK 808
Volcanic ash condition	055 050 00	05-51 TASK 803
21-0 AIR CONDITIONING - GENERAL		
Smoke		
• does not stop with recirculation fans off, isolation valve closed, and both pack switches at off		
- flight compartment	210 070 00	21-00 TASK 803
- flight compartment and passenger cabin	210 080 00	21-00 TASK 803
- passenger cabin	210 090 00	21-00 TASK 803
• stops with recirculation fan(s) off, isolation valve closed, and right pack switch at off		
- flight compartment	210 010 00	21-00 TASK 801
- flight compartment and passenger cabin	210 020 00	21-00 TASK 801
- passenger cabin	210 030 00	21-00 TASK 801

EFFECTIVITY
XXX ALL

OBSERVED FAULT LIST
SYSTEM-ORDER

图 3-13　可观察故障列表：按系统索引

（13）客舱故障列表（Cabin Fault List）：客舱故障清单（分散在各系统中）包括通话 / 公共地址、门 / 窗、应急设备、厨房、娱乐设备、灯光、各种各样的性质、坐椅、厕所 / 马桶等，如图 3-14 所示。

BOEING®
737-600/700/800/900
FAULT ISOLATION MANUAL

FAULT DESCRIPTION	FAULT CODE (location - - -)	GO TO FIM TASK
COMMUNICATIONS/PUBLIC ADDRESS		
Cabin attendant handset:		
• damaged	C12 12 - - -	AIRLINE METHOD
• distorted	C12 20 - - -	23-41 TASK 804
• inoperative	C12 33 - - -	23-41 TASK 804
• volume problem	C12 65 - - -	23-41 TASK 804
Chimes:		
• does not turn off	C14 83 - - -	23-31 TASK 803
• does not turn on	C14 84 - - -	23-31 TASK 804
• volume problem	C14 65 - - -	23-31 TASK 802
Chimes light:		
• does not turn off	C18 83 - - -	33-20 TASK 801
• does not turn on	C18 84 - - -	33-20 TASK 801
Passenger address system:		
• distorted	C16 20 - - -	23-31 TASK 805
• inoperative	C16 33 - - -	23-31 TASK 805
• intermittent	C16 34 - - -	23-31 TASK 805
• volume problem	C16 65 - - -	23-31 TASK 802
Speaker:		
• inoperative	C17 33 - - -	23-31 TASK 806
• intermittent	C17 34 - - -	23-31 TASK 806
• noisy	C17 42 - - -	23-31 TASK 806
• volume problem	C17 65 - - -	23-31 TASK 806

EFFECTIVITY
XXX ALL

CABIN FAULT LIST

图 3-14　客舱故障列表

（14）客舱故障定位（Cabin Fault Locator）：提供客舱故障区域及位置，如图 3-15 所示。

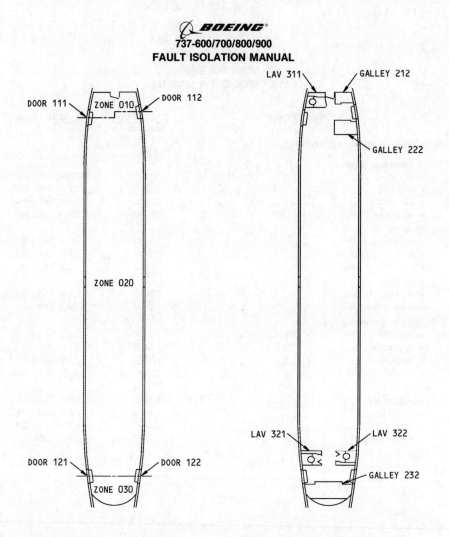

图 3-15　客舱故障定位

（15）客舱故障代码索引（Cabin Fault Code Index）：客舱故障代码包含数字与字母且顺序表示，如图 3-16 所示。

737-600/700/800/900
FAULT ISOLATION MANUAL

FAULT CODE	FAULT DESCRIPTION	GO TO FIM TASK
C12 12 - - -	Cabin attendant handset: damaged.	AIRLINE METHOD
C12 20 - - -	Cabin attendant handset: distorted.	23-41 TASK 804
C12 33 - - -	Cabin attendant handset: inoperative.	23-41 TASK 804
C12 65 - - -	Cabin attendant handset: volume problem.	23-41 TASK 804
C14 65 - - -	Chimes: volume problem.	23-31 TASK 802
C14 83 - - -	Chimes: does not turn off.	23-31 TASK 803
C14 84 - - -	Chimes: does not turn on.	23-31 TASK 804
C16 20 - - -	Passenger address system: distorted.	23-31 TASK 805
C16 33 - - -	Passenger address system: inoperative.	23-31 TASK 805
C16 34 - - -	Passenger address system: intermittent.	23-31 TASK 805
C16 65 - - -	Passenger address system: volume problem.	23-31 TASK 805
C17 33 - - -	Speaker: inoperative.	23-31 TASK 806
C17 34 - - -	Speaker: intermittent.	23-31 TASK 806
C17 42 - - -	Speaker: noisy.	23-31 TASK 806
C17 65 - - -	Speaker: volume problem.	23-31 TASK 806
C18 83 - - -	Chimes light: does not turn off.	33-20 TASK 801
C18 84 - - -	Chimes light: does not turn on.	33-20 TASK 801
D11 15 - - -	Cabin window: cracked.	AIRLINE METHOD
D11 17 - - -	Cabin window: dirty.	AIRLINE METHOD
D11 24 - - -	Cabin window: fogged/moisture between panes.	AIRLINE METHOD
D11 36 - - -	Cabin window: leaking.	AIRLINE METHOD
D11 42 - - -	Cabin window: noisy.	AIRLINE METHOD
D11 51 - - -	Cabin window: scratched.	AIRLINE METHOD
D12 12 - - -	Cabin window shade: damaged.	AIRLINE METHOD
D12 29 - - -	Cabin window shade: hard to operate.	25-21 TASK 803
D12 37 - - -	Cabin window shade: loose.	25-21 TASK 803
D12 39 - - -	Cabin window shade: missing.	AIRLINE METHOD

EFFECTIVITY
XXX ALL

CABIN FAULT CODE INDEX

图 3-16 客舱故障代码索引

任务实施

1. 飞机有效性查询

打开前言（Front Matter），单击飞机有效性列表（Effective Aircraft）查询飞机有效性代码，如图 3-17 所示。

737-600/700/800/900
FAULT ISOLATION MANUAL

This manual is applicable to the aircraft in this list:

| Model-Series | Operator | | Manufacturer | | | Registration Number |
	Identification Code	Effectivity Code	Block Number	Serial Number	Line Number	
737-7W0	XXX	001	YA811	29912	140	B-2639
737-7W0	XXX	002	YA812	29913	148	B-2640
737-7W0	XXX	003	YA813	30074	292	B-2503
737-7W0	XXX	004	YA814	30075	311	B-2502

图 3-17　飞机有效性列表

2. 读懂故障隔离手册的飞机有效性列表

与项目 1 中提到的波音飞机维护手册类似，由图 3-17 我们可以轻松判断飞机的有效性代码、造册号等信息。

思考题

1. 从飞机的有效性列表来看，AMM 与 FIM 很相似，那么在使用 FIM 时能不能将 AMM 的有效性代替呢？

2. 相较前面我们学习过的 AMM 与 IPC，你能说出 FIM 的前言部分与它们的异同点吗？

任务 2　飞机防滞/自动刹车控制器章节查询

任务描述

故障隔离手册的有效性查询确认了飞机的有效性和适用性。接下来，我们将使用满足飞机有效性的 FIM 手册查询并明确飞机防滞/自动刹车控制器所在章节。

故障隔离手册（FIM）与零部件图解目录手册（IPC）、飞机维护手册（AMM）都属于客户化手册，章节划分按照《ATA 100 规范》编排。我们只要充分熟悉《ATA 100 规范》章节划分的原则，就能快速地查询并确定飞机防滞/自动刹车控制器所在章。

你还记得《ATA 100 规范》中，飞机防滞/自动刹车控制器归属哪个系统吗？借助已学过的 AMM PART I SDS（Systems Description Section），你是否能获取防滞/自动刹车控制器的相关信息？

任务要求

● 了解故障隔离手册（FIM）的结构。

● 了解飞机防滞 / 自动刹车控制器相关信息的查询。

知识链接

1. FIM 的结构

FIM 手册的结构和 AMM 基本相同，由前言和各系统章节组成。前言（Front Matter）除包含了本册的有效页清单、修订改版说明、服务通告（SB），还包括了可观察故障清单、客舱故障清单等，该内容在任务 1 中已详细附图介绍，该处不再赘述。FIM 手册各系统目录包括两部分：前部内容和故障隔离工卡。如图 3-18 所示，前部内容包含有效页清单、手册使用、故障代码索引、维护信息索引。

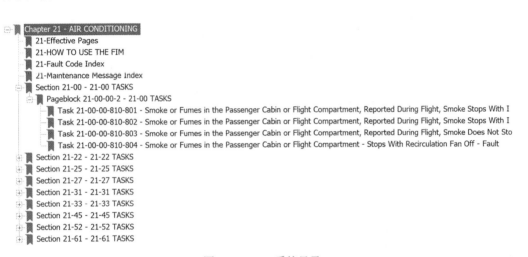

图 3-18　FIM 系统目录

各系统章节的故障代码索引是该章节所有故障代码，按代码数字顺序排列（故障代码来自故障报告手册，每一个故障都对应 8 位阿拉伯数字代码），如图 3-19 所示，在故障代码索引中，每一个代码都有相应的故障描述和故障隔离手册的工卡号，以便按照代码查询到故障隔离工卡，实际工作中也可以根据代码索引里的故障描述直接查找工卡。

同样，各系统的维护信息索引都按照部件字母顺序排列，列出了该系统章节所有相关的维护信息，对于自检有代码的系统或部件，7 位故障代码同样显示在维护信息内容中，如图 3-19 所示。

2. 飞机防滞 / 自动刹车系统的相关知识

飞机防滞 / 自动刹车系统作为飞机重要的机载设备，对飞机的起飞、安全着陆、持续作战和适应机场的能力等方面起着重要的作用。刹车系统性能的好坏直接影响到飞机及机载人员的安全。由于飞机着陆过程持续的时间比较短，因此要求飞机防滞 / 自动刹车系统在飞机着陆后或中断起飞过程中，必须安全、反应迅速，以确保飞机安全刹车。该系统主要包含的部件有 AACU（防滞 / 自动刹车控制组件）、自

飞机的安全刹车

动刹车压力控制组件、两个自动刹车梭形活门、自动刹车选择控制面板。AACU 控制组件接收速度刹车预位电门、自动油门电门组件、ADIRU、PSEU、自动刹车压力控制组件、自动刹车梭形活门、防滞传感器、起落架手柄收上位电门等输入信息控制自动刹车系统工作。

BOEING
737-600/700/800/900
FAULT ISOLATION MANUAL

FAULT CODE	FAULT DESCRIPTION	GO TO FIM TASK
210 010 00	Smoke: stops with recirculation fan(s) off, isolation valve closed, and right pack switch at off - flight compartment.	21-00 TASK 801
210 020 00	Smoke: stops with recirculation fan(s) off, isolation valve closed, and right pack switch at off - flight compartment and passenger cabin.	21-00 TASK 801
210 030 00	Smoke: stops with recirculation fan(s) off, isolation valve closed, and right pack switch at off - passenger cabin.	21-00 TASK 801
210 040 00	Smoke: stops with recirculation fan(s) off, isolation valve closed, right pack switch in auto, and left pack switch off - flight compartment.	21-00 TASK 802
210 050 00	Smoke: stops with recirculation fan(s) off, isolation valve closed, right pack switch in auto, and left pack switch off - flight compartment and passenger cabin.	21-00 TASK 802
210 060 00	Smoke: stops with recirculation fan(s) off, isolation valve closed, right pack switch in auto, and left pack switch off - passenger cabin.	21-00 TASK 802
210 070 00	Smoke: does not stop with recirculation fans off, isolation valve closed, and both pack switches at off - flight compartment.	21-00 TASK 803
210 080 00	Smoke: does not stop with recirculation fans off, isolation valve closed, and both pack switches at off - flight compartment and passenger cabin.	21-00 TASK 803
210 090 00	Smoke: does not stop with recirculation fans off, isolation valve closed, and both pack switches at off - passenger cabin.	21-00 TASK 803
212 010 31	Windshield: air outlet does not operate - captain's.	21-22 TASK 801
212 010 32	Windshield: air outlet does not operate - first officer's.	21-22 TASK 801
212 020 31	Foot air outlet: does not operate - captain's.	21-22 TASK 801
212 020 32	Foot air outlet: does not operate - first officer's.	21-22 TASK 801
212 040 00	Recirculation fan (single-fan system): does not operate.	21-25 TASK 801
212 050 00	Smoke: stops with recirculation fan(s) off - flight compartment.	21-00 TASK 804
212 060 00	Smoke: stops with recirculation fan(s) off - flight compartment and passenger cabin.	21-00 TASK 804
212 070 00	Smoke: stops with recirculation fan(s) off - passenger cabin.	21-00 TASK 804
212 080 00	EQUIP COOLING EXHAUST OFF light: light on with switch at ALTERNATE.	21-27 TASK 805
212 090 00	EQUIP COOLING EXHAUST OFF light: light on with switch at NORMAL.	21-27 TASK 806
212 100 00	EQUIP COOLING SUPPLY OFF light: light on with switch at ALTERNATE.	21-27 TASK 807
212 110 00	EQUIP COOLING SUPPLY OFF light: light on with switch at NORMAL.	21-27 TASK 808

EFFECTIVITY
XXX ALL

21-FAULT CODE INDEX

图 3-19　故障代码索引

737-600/700/800/900
FAULT ISOLATION MANUAL

LRU/SYSTEM	MAINTENANCE MESSAGE	GO TO FIM TASK
35 DEG CONT L	POSN 1 - NO GO	21-52 TASK 802
35 DEG CONT L	POSN 2 - GO, 35F CONTROL VALVE NOT OPEN	21-52 TASK 812
35 DEG CONT L	POSN 2 - NO GO	21-52 TASK 804
35 DEG CONT L	POSN 3 - NO GO	21-52 TASK 806
35 DEG CONT L	POSN 4 - GO, 35F CONTROL VALVE NOT CLOSED	21-52 TASK 813
35 DEG CONT L	POSN 4 - NO GO	21-52 TASK 808
35 DEG CONT L	POSN 5 - NO GO	21-52 TASK 810
35 DEG CONT R	POSN 1 - NO GO	21-52 TASK 803
35 DEG CONT R	POSN 2 - GO, 35F CONTROL VALVE NOT OPEN	21-52 TASK 814
35 DEG CONT R	POSN 2 - NO GO	21-52 TASK 805
35 DEG CONT R	POSN 3 - NO GO	21-52 TASK 807
35 DEG CONT R	POSN 4 - GO, 35F CONTROL VALVE NOT CLOSED	21-52 TASK 815
35 DEG CONT R	POSN 4 - NO GO	21-52 TASK 809
35 DEG CONT R	POSN 5 - NO GO	21-52 TASK 811
CAB PRESS CON	ACFT RATE HI	21-31 TASK 815
CAB PRESS CON	AUTO/MAN ERROR	21-31 TASK 821
CAB PRESS CON	CAB DIFF PRESS HI	21-31 TASK 816
CAB PRESS CON	CAB PRESS SW ACTIV	21-31 TASK 820
CAB PRESS CON	CAB RATE HI	21-31 TASK 818
CAB PRESS CON	CONTRLR LRU FAIL	21-31 TASK 801
CAB PRESS CON	DADC #1 INVALID	21-31 TASK 806
CAB PRESS CON	DADC #2 INVALID	21-31 TASK 807
CAB PRESS CON	DRV ENBL ERROR	21-31 TASK 819
CAB PRESS CON	HI CAB ALT	21-31 TASK 817
CAB PRESS CON	LDG GEAR DISAGREE	21-31 TASK 822
CAB PRESS CON	LO INFL/HI LEAKG	21-31 TASK 823
CAB PRESS CON	N1/N2 LOWRU FAIL	21-31 TASK 810
CAB PRESS CON	OFV LRU FAIL	21-31 TASK 802

EFFECTIVITY
XXX ALL

21-MAINT MSG INDEX

图 3-20　维护信息索引

　　自动刹车系统监控机轮减速并控制刹车计量压力与维持驾驶员在自动刹车（AUTO BRAKE）选拔电门上选定的压力一致，直到飞机完全停住。自动刹车功能包括：①挡 1、挡 2、挡 3 和最大（MAX）减速位置在着陆刹车过程中控制自动刹车系统来调整压力直到飞机完全停住；②中断起飞（RTO）位置控制自动刹车系统提供全部压力到机轮刹车并使飞机停住。当驾驶员在地速超过 88 节开始中断起飞时自动刹车系统工作在 RTO 模式。

　　民用飞机着陆前，首先要在空中对准跑道，然后向后收油门，减少发动机的推力。

当飞机开始减速，速度降低到允许值以下时，放下起落架。然后，放下部分安装在主机翼上的副翼（不能完全放到底），使飞机下降；并向后拉驾驶杆（有的机型是驾驶盘），使尾翼向上偏转，飞机机头略微抬起。飞行员要保持这种飞行姿态，使飞机逐渐下滑，并不断修正方向和下滑角度的误差，直到飞机的主起落架接触跑道；此时，飞行员要把副翼完全放下（空气阻力增加到最大值），并向前推驾驶杆（驾驶盘）使尾翼改平，使机头放平，前起落架接触地面。飞行员踩下起落架的刹车，增加轮胎转动的阻力，使飞机逐渐减速。有些机型的发动机有反推力装置，能改变喷气方向，使发动机向后喷出的气流转向前方，帮助飞机减速。综上，在飞机接触地面以后，减速主要依靠起落架刹车、放下副翼（不是尾翼）、发动机反推力装置（部分机型）。另外，军用作战飞机还有两种民航机没有的减速措施：减速板和减速伞。

💬 任务实施

1. 飞机防滞/自动刹车器相关信息查询

对于飞机防滞/自动刹车器相关信息，可查询满足飞机有效性的 AMM PART I SDS（Systems Description Section）32 章起落架系统（Landing Gear）- 车轮和刹车（WHEELS AND BRAKES）。如图 3-21 所示，车轮和刹车系统包括轮胎和车轮、液压制动系统、停车制动系统、防滞/自动刹车系统。

图 3-21 飞机自动刹车 SDS 查询

2. 飞机防滞 / 自动刹车器章节查询

打开 FIM 手册前言部分的可观察故障清单，搜索自动刹车（Auto Brake），确定飞机防滞 / 自动刹车器所在章节，如图 3-22 所示。

飞机防滞 / 自动刹车器
章节查询

BOEING®

737-600/700/800/900

FAULT ISOLATION MANUAL

FAULT DESCRIPTION	FAULT CODE	GO TO FIM TASK
ATC (continued)		
• code indicator does not respond to selector movement . . .	345 120 00	34-53 TASK 814
• transmission weak or intermittent		
- no. 1.	345 130 01	34-53 TASK 802
- no. 1 and no. 2	345 130 48	34-53 TASK 801
- no. 2.	345 130 02	34-53 TASK 803
• transponder fail light on	345 140 00	34-53 TASK 801
ATT flag shows on the EFIS display		
- captain's	342 010 31	34-21 TASK 827
- first officer's	342 010 32	34-21 TASK 828
Attitude display		
• does not show		
- captain's	342 015 31	34-21 TASK 827
- first officer's	342 015 32	34-21 TASK 828
Audio control panel indicator light problem		
- captain's	235 101 31	23-51 TASK 807
- first observer's	235 101 33	23-51 TASK 807
- first officer's	235 101 32	23-51 TASK 807
Audio control panel selector switch problem		
- captain's	235 103 31	23-51 TASK 807
- first observer's	235 103 33	23-51 TASK 807
- first officer's	235 103 32	23-51 TASK 807
Audio control panel volume control problem		
- captain's	235 102 31	23-51 TASK 807
- first observer's	235 102 33	23-51 TASK 807
- first officer's	235 102 32	23-51 TASK 807
Auto brake		
• deceleration rate is not correct	324 090 00	32-42 TASK 807
AUTO BRAKE DISARM light		
• light on	324 080 00	32-42 TASK 828
AUTO FAIL light		
• light on and ALTN light on, AUTO FAIL light goes off when pressurization mode selector switch is at ALTN	213 010 00	21-31 TASK 801
• light on, ALTN light not on	213 020 00	21-31 TASK 801
AUTO SLAT FAIL light		
• light on	278 060 00	27-32 TASK 801

EFFECTIVITY

XXX **ALL**

OBSERVED FAULT LIST

ALPHABETICAL

图 3-22　FIM 3 手册前言部分

因此，可确认飞机防滞/自动刹车器故障隔离相关信息在 32 章。

思考题

1. 如果 APU 故障，你知道应该如何确定章节吗？
2. 可观察故障清单中对应的工卡是否能直接对应该故障隔离工卡？

任务 3　飞机防滞/自动刹车器故障隔离

任务描述

已知飞机防滞/自动刹车器故障的维护信息（MAINTENANCE MESSAGE）为 AIR/GND R584 FLT（或维护信息代码为 32-06001），需确定该故障指向的故障任务号（GO TO FIM TASK）。上一任务中，我们已经知晓了飞机防滞/自动刹车器故障所在的章节。为了制定飞机防滞/自动刹车器故障隔离方案，我们还需要深入探寻飞机防滞/自动刹车器故障任务卡中故障产生可能的原因（Possible Causes）、跳开关（Circuit Breakers）、相关资料（Related Data）、初始评估（Initial Evaluation）、故障隔离程序（Fault Isolation Procedure）等信息。

如何利用 FIM 手册查询故障最可能的原因、相关跳开关设备号及位置？如何在 if… then… 的句式中快速找到故障隔离的窍门？

任务要求

● 掌握 FIM 手册中自检维护信息的使用。
● 掌握故障隔离程序工卡的编制。

知识链接

两种方法查询 FIM 手册
维护信息

1. FIM 手册维护信息

维护信息（Maintenance Message）是通过机内嵌入式自检设备（Built-in Test Equipment，BITE）得到的飞机某系统或部件的故障指示。它有助于发现一个可观察故障的原因。一般维护信息包括特殊的灯或灯亮、一组文字代码、数字代码。维护信息有两种方式可查询：①在每章中都有维护信息索引表（图 3-20）；②通过可观察故障清单或客舱故障清单找到对应的章节，可在自检程序（BITE Procedure）完成后找到对应章节的维护信息列表（图 3-23）。

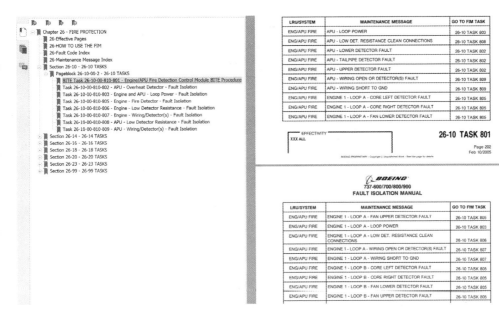

图 3-23　26-10 维护信息列表示意图

无论是哪一种方式，只要查询到与故障完全一致的维护信息或维护信息代码，即可根据维护信息列表中对应的故障任务号，找到对应的故障隔离工卡。

2．故障隔离工卡的阅读

故障隔离工卡中通常主要包括故障概述（General）、故障描述（Description）、故障可能的原因（Possible Causes）、跳开关信息（Circuit Breakers）、相关资料（Related Data）、初始评估（Initial Evaluation）、故障隔离程序（Fault Isolation Procedure）等信息。对于不同的故障，我们只有深入探究了这些相关信息，才能科学地制定专属的故障隔离工卡。

A．故障概述（General）：故障隔离手册是一个帮你修理飞机的工具。在故障隔离手册里的程序可以帮你更快隔离每一架飞机的故障原因。隔离故障可以用你知道的方法、过去的排故经验、故障发生的条件、飞机或机群的故障历史。不强制要求用故障隔离程序步骤来排故。当不能确定计划时，需在排故前阅读故障隔离手册程序，按照故障隔离程序完成故障隔离步骤，如图 3-24 所示。

A．故障描述（Description）：描述段落给出故障状态下的故障信息，描述故障的原因、条件、输入信号错误，系统详细信息参照 AMM Part I，如图 3-25 所示。

B．故障可能的原因（Possible Causes）：通过可能原因清单可快速对应故障隔离程序里的故障原因，并且这些原因可通过故障隔离程序逐一排查。注意在执行隔离程序前，先查询航空公司的维修记录，以确认哪些原因是已经排查过的。可能原因清单采用排序列举法编制，最可能的原因在第一个，最不可能的原因在最后，如图 3-26 所示。

C．跳开关信息（Circuit Breakers）：跳开关清单给出故障涉及的相关跳开关信息，包括跳开关名称、跳开关所在的面板和位置。例如：如图 3-27 所示，设备号为 C00403 的 APU 消防检测器在跳开关面板 P6-2 的 A 行、23 列。

737-600/700/800/900

FAULT ISOLATION MANUAL

801. **Engine/APU Fire Detection Control Module BITE Procedure**

  A. General

    (1) The engine & APU fire detection control module, M279 is located on the E2-2 shelf in the electronic equipment (EE) compartment. The engine & APU fire detection control module will be referred to as the control module throughout this procedure. Access the front panel of the control module to do the BITE test.

    (2) The front of the control module contains five amber fault area lights, a FAULT/INOP TEST switch, and three red fault display lights. The control module has these fault lights:

      (a) Engine 1 Loop A fault area light

      (b) Engine 1 Loop B fault area light

      (c) Engine 2 Loop A fault area light

      (d) Engine 2 Loop B fault area light

      (e) APU fault area light

      (f) Three fault area lights which indicate the type of fault and the approximate location of the fault, if it is known.

    (3) The amber fault area lights indicate which loop has a fault. The red fault display lights tell the type of fault and location, if known. If there is a fault, the applicable fault area lights will stay on until the fault is corrected. For each combination of fault display lights, there is a related maintenance message.

    (4) If more than one of the amber fault area lights come on, then there are multiple faults. In this case, the control module employs a priority sequence. The red fault display lights indicate the fault for the first loop in the sequence. When that fault is corrected, the corresponding amber light goes off, and the red fault display lights change to indicate the fault for the next loop in the sequence. The priority sequence follows:

      (a) Engine 1 Loop A

      (b) Engine 1 Loop B

      (c) Engine 2 Loop A

      (d) Engine 2 Loop B

      (e) APU

    (5) To do the BITE test, push and hold the FAULT/INOP TEST switch for five seconds. If all of the fault lights come on when you do the test, and all of the fault lights go off when you release the switch, the test passes. If fault lights stay on after you release the switch, there is a fault.

图 3-24　故障概述示例

737-600/700/800/900

FAULT ISOLATION MANUAL

802. **APU - Overheat Detector - Fault Isolation**

  A. Description

    (1) This task is for these maintenance messages:

      (a) APU - UPPER

      (b) APU - LOWER

      (c) APU - TAILPIPE

    (2) These messages occur when the control module, M279 detects a fault with an APU fire detector. The detectors are listed in the table below.

DESCRIPTION	EQUIPMENT NUMBER
APU UPPER OVERHEAT DETECTOR ASSEMBLY	M1755
APU LOWER OVERHEAT DETECTOR ASSEMBLY	M1756
APU TAILPIPE OVERHEAT DETECTOR ASSEMBLY	M1925

图 3-25　故障描述示例

B. Possible Causes

    (1) Fire detector Element

    (2) Control module, M279.

图 3-26　故障可能的原因示例

C. Circuit Breakers

    (1) This is the primary circuit breaker related to the fault:

        F/O Electrical System Panel, P6-2

Row	Col	Number	Name
A	23	C00403	FIRE PROTECTION DETECTION APU

图 3-27　跳开关信息示例

D. 相关资料（Related Data）：该段落给出可参考的其他手册资料去解决故障，如系统原理图手册（System Schematics Manual，SSM）和线路图手册（Wiring Diagram Manual，WDM），如图 3-28 所示。

D. Related Data

    (1) (SSM 26-00-01)

    (2) (SSM 26-11-31)

    (3) (WDM 26-11-31)

图 3-28　相关资料示例

E. 初始评估（Initial Evalution）：多数故障隔离程序中都有该部分，它应用于故障条件依然存在的情况，有利于快速地检查和了解故障。如果初检不能确认故障，则可判定为间歇性故障，此时依照所在公司的意见和维护记录去处理，如图 3-29 所示。

E. Initial Evaluation

    (1) Do this task: Engine/APU Fire Detection Control Module BITE Procedure, 26-10 TASK 801.

        (a) If a maintenance message shows, then do the Fault Isolation Procedure below.

        (b) If the maintenance message does not show, then there was an intermittent fault.

图 3-29　初始评估示例

对于一些显而易见的故障（如雨刮丢失、排水管结冰和天花板受潮），或者地面无法再现的故障（如发动机喘振、自动驾驶故障），隔离程序中没有涉及。部分或所有的初始评估，是去设置一些初始条件或者找出是哪个故障隔离工作程序是必需的。

F. 故障隔离程序（Fault Isolation Procedure）：故障隔离程序提供若干个轨道步骤的顺序做故障隔离。如图 3-30 所示，程序前面的步骤是为故障隔离程序做准备，最后的步骤则是让飞机恢复正常。隔离程序通常从快速检查开始，如未能确认故障则按需进行系统专项测试以确认故障可能的原因。

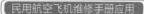

F. Fault Isolation Procedure

(1) Prepare the airplane for fault isolation.

  (a) Open these circuit breakers and install safety tags:

F/O Electrical System Panel, P6-2

Row	Col	Number	Name
A	23	C00403	FIRE PROTECTION DETECTION APU
B	19	C01344	APU FIRE SW POWER

F/O Electrical System Panel, P6-4

Row	Col	Number	Name
A	14	C00033	AUX POWER UNIT CONT

  (b) To access the upper or lower APU detector, open this access panel:

Number	Name/Location
315A	APU Cowl Door

  (c) To access the APU tailpipe overheat detector, open this access panel:

Number	Name/Location
318BR	Tailcone Access Door

(2) Replace the detector element called out in the maintenance message. These are the tasks:
- APU Overheat Detector Element Removal, AMM TASK 26-15-01-000-801
- APU Overheat Detector Element Installation, AMM TASK 26-15-01-400-801

  (a) If the replacement test passes, then you corrected the fault.

  (b) If the replacement test fails, then continue.

(3) Replace the control module, M279. These are the tasks:
- Engine and APU Fire Detection Module Removal, AMM TASK 26-10-01-000-801
- Engine and APU Fire Detection Module Installation, AMM TASK 26-10-01-400-801

  (a) If the replacement test passes, then you corrected the fault.

(4) Return the airplane to its usual condition.

Close this access panel:

Number	Name/Location
315A	APU Cowl Door

图 3-30　故障隔离程序示例

鉴于故障隔离程序的复杂性，归纳三条使用规则：

（1）按故障隔离程序指定的顺序步骤做故障隔离。

（2）遇到"如果……然后……"（if…then…）的说明，需要放慢阅读速度，因为这将引导我们正确隔离故障。

（3）完成所有的步骤，意味着我们完成了故障隔离。

任务实施

1. 飞机防滞／自动刹车器自检维护信息查询

本项目任务 2 已确定飞机防滞／自动刹车器在 32 章，因此，我们可以打开 FIM 32 章的维护信息索引总表（Maintenance Message Index），搜索维护信息 AIR/GND R584 FLT 或维护信息代码 32-06001。

如图 3-31 所示，飞机防滞／自动刹车控制器故障任务号为 32-09 TASK 804。

737-600/700/800/900
FAULT ISOLATION MANUAL

LRU/SYSTEM	MAINTENANCE MESSAGE	GO TO FIM TASK
PSEU	32-01007 AIR/GND FAIL	32-09 TASK 819
PSEU	32-02001 PARK BRK FAULT	32-09 TASK 806
PSEU	32-02002 PARK BRK A FAULT	32-09 TASK 806
PSEU	32-02003 PARK BRK B FAULT	32-09 TASK 806
PSEU	32-03001 PARK BRK A SET	32-09 TASK 806
PSEU	32-03002 PARK BRK B SET	32-09 TASK 806
PSEU	32-04001 AIR/GND OVERRIDE	32-09 TASK 818
PSEU	32-06001 AIR/GND R584 FLT	32-09 TASK 804
PSEU	32-06002 AIR/GND R593 FLT	32-09 TASK 804
PSEU	32-06003 AIR/GND R587 FLT	32-09 TASK 804
PSEU	32-06004 AIR/GND R583 FLT	32-09 TASK 804
PSEU	32-06005 AIR/GND R589 FLT	32-09 TASK 804
PSEU	32-06006 AIR/GND R592 FLT	32-09 TASK 804
PSEU	32-06007 AIR/GND R594 FLT	32-09 TASK 804
PSEU	32-06008 LG LVR LCH FLT	32-09 TASK 807
PSEU	32-06009 AIR/GND R585 FLT	32-09 TASK 804
PSEU	32-06010 AIR/GND R588 FLT	32-09 TASK 804
PSEU	32-06011 AIR/GND R595 FLT	32-09 TASK 804
PSEU	32-06012 AIR/GND R586 FLT	32-09 TASK 804
PSEU	32-06013 AIR/GND R590 FLT	32-09 TASK 804
PSEU	32-06014 AIR/GND R591 FLT	32-09 TASK 804
PSEU	32-06015 AIR/GND R597 FLT	32-09 TASK 804
PSEU	32-06016 NGS RLY FAULT	32-61 TASK 813
PSEU	32-31001 L ON GND A FAULT	32-09 TASK 803
PSEU	32-31002 N ON GND A FAULT	32-09 TASK 802
PSEU	32-31003 R ON GND A FAULT	32-09 TASK 803
PSEU	32-31004 L ON GND B FAULT	32-09 TASK 803
PSEU	32-31005 N ON GND B FAULT	32-09 TASK 802

EFFECTIVITY
XXX ALL

32-MAINT MSG INDEX

图 3-31 32 章维护信息索引表

2. 飞机防滞 / 自动刹车控制器故障隔离

打开 FIM 32-09 TASK 804，阅读工卡中的概述、故障可能的原因、跳开关信息、初始评估、故障隔离程序等信息。

飞机防滞 / 自动刹车控制器
故障隔离工卡的阅读

如图 3-32 所示，飞机防滞 / 自动刹车控制器故障维护信息为 AIR/GND R584 FLT 时，最可能的故障原因是自动刹车系统 1 中空 / 地继电器 R584、R587、R589、R592、R593、R594 故障，其次是自动刹车系统 2 中空 / 地继电器 R585、R588、R590、R591、R595、R596 故障，再次是线路故障，最不可能的故障原因是临近电门电子组件（Proximity switch electronics unit，PSEU）M2061 故障。

BOEING®
737-600/700/800/900
FAULT ISOLATION MANUAL

  • Proximity Switch Electronics Unit (PSEU) Installation, AMM TASK 32-09-10-400-801

  (b) Do the post installation test in the PSEU installation procedure.

  (c) If the test operates correctly, then you corrected the fault.

G. Repair Confirmation

  (1) Do this test of the applicable air/ground sensor:

  (a) Put a deactuator on the face of the sensor face.

  (b) Remove the deactuator from the face of the sensor face.

  (c) Do the EXISTING FAULTS test on the PSEU BITE display. Do this task: Proximity Switch Electronics Unit (PSEU) BITE Procedure, 32-09 TASK 801.

  1) If you do not find the maintenance message, then you corrected the fault.

————————— END OF TASK —————————

804. Air/Ground Relay Fault - Fault Isolation

  A. Description

  (1) This task is for these maintenance messages:

  (a) 32-06001 AIR/GND R584 FLT

  (b) 32-06002 AIR/GND R593 FLT

  (c) 32-06003 AIR/GND R587 FLT

  (d) 32-06004 AIR/GND R583 FLT

  (e) 32-06005 AIR/GND R589 FLT

  (f) 32-06006 AIR/GND R592 FLT

  (g) 32-06007 AIR/GND R594 FLT

  (h) 32-06009 AIR/GND R585 FLT

  (i) 32-06010 AIR/GND R588 FLT

  (j) 32-06011 AIR/GND R595 FLT

  (k) 32-06012 AIR/GND R586 FLT

  (l) 32-06013 AIR/GND R590 FLT

  (m) 32-06014 AIR/GND R591 FLT

  (n) 32-06015 AIR/GND R597 FLT

  (2) These maintenance messages show that there is a problem with a air/ground relay.

  (a) The PSEU does not sense the expected load from the applicable air/ground relay.

B. Possible Causes

  (1) System 1 air/ground relay, R584, R587, R589, R592, R593, or R594

  (2) System 2 air/ground relay, R585, R588, R590, R591, R595, or R596

  (3) Wiring problem

  (4) Proximity switch electronics unit (PSEU), M2061

EFFECTIVITY
XXX ALL

32-09 TASKS 803-804

图 3-32 防滞 / 自动刹车故障原因

　　继续 32-09 TASK 804 工卡的阅读，如图 3-33 所示，我们可以知晓，防滞 / 自动刹车控制器故障相关跳开关位置在副驾驶电气系统面板（F/O Electrical System Pane）P6-3。其中，起落架空 / 地系统 2 中的跳开关 C01355 在面板 P6-3 的 C 行 15 列，起落架空 / 地继电器的跳开关 C01401 在面板 P6-3 的 D 行 15 列。初始评估中首先提到，需要确保这两个跳开关处于闭合状态。

BOEING

737-600/700/800/900

FAULT ISOLATION MANUAL

C. Circuit Breakers

　(1) These are the primary circuit breakers related to the fault:

　　F/O Electrical System Panel, P6-3

Row	Col	Number	Name
C	15	C01355	LANDING GEAR AIR/GND SYS 2
D	15	C01401	LANDING GEAR AIR/GND RELAY

D. Related Data

　(1) (SSM 32-09-11)

　(2) (SSM 32-09-12)

　(3) (WDM 32-31-11)

　(4) (WDM 32-31-12)

E. Initial Evaluation

　(1) Make sure that these circuit breakers are closed:

　　F/O Electrical System Panel, P6-3

Row	Col	Number	Name
C	15	C01355	LANDING GEAR AIR/GND SYS 2
D	15	C01401	LANDING GEAR AIR/GND RELAY

　(a) If a circuit breaker opens, continue to troubleshoot.

　(b) If the circuit breaker was open and stays closed, do the Repair Confirmation steps at the end of the procedure.

　(2) Make sure the relay is installed tightly in the J20 or J22 panel.

　(a) If the relay is loose, then do these steps:

　　1) Tighten the relay.

　　2) Do the Repair Confirmation at the end of this task.

　(b) If the relay is secure, then do the Fault Isolation Procedure below.

F. Fault Isolation Procedure

　(1) Do this check for power to the applicable relay:

　(a) Remove the applicable relay.

　(b) Do a check for 28 VDC between pin X1 of the relay socket and structure ground.

　　1) If there is not 28 VDC at pin X1, then do these steps:

　　　a) Repair the wiring between the circuit breaker and relay.

　　　b) Re-install the relay.

　　　c) Do the Repair Confirmation at the end of this task.

　　2) If there is 28 VDC at pin X1, then continue.

　(2) Install a new relay.

　(a) Do the Repair Confirmation at the end of this task.

　　1) If the Repair Confirmation is not satisfactory, then continue.

　(3) Do this check of the wiring:

EFFECTIVITY

XXX **ALL**

32-09 TASK 804

图 3-33　防滞 / 自动刹车故障隔离程序

阅读故障隔离程序段落时，注意"如果引脚 X1 没有 28V 直流电压，我们需要做以下步骤"（If there is not 28VDC at pin X1, then do these steps）的句子，当实际故障检测满足这条时，就可以通过接下来的步骤完成故障维修；如果不满足，需要按照手册继续做故障隔离，如工卡中的第（2）步安装新的继电器（Install a new relay），如果维修效果不理想，则进行第（3）步线路检测（Do this check of the wiring）……依次进行，直到完成故障维修。

思考题

1．如果飞机防滞 / 自动刹车控制器自检维护信息为 ADIRU L，你能模仿该项目的查询方法确定故障隔离工卡号吗？

2．可观察故障清单中对应的故障隔离工卡中一定能找到相适应的故障隔离程序吗？

工卡实操

本项目配套民航机务维修典型案例——FIM 应用工卡 6 个，专业面向包括 AV 和 ME，难度系数设置了Ⅲ（易）、Ⅱ（中）、Ⅰ（难）三个等级，学习者可按需选用。工卡实操过程中，学习者不仅需要关注修理任务本身，同时需有意识地查询手册中相关任务的保障人身与设备安全的警告（Warning）和注意事项（Caution/Note）。工作中，工作者需要确认已经查询或知晓了相关注意事项和警告，并在工卡上签署确认，才可开始相应任务。本项目配套工卡详情见表 3-1。

表 3-1　项目 3 配套工卡列表

工卡序号	项目名称	专业方向	难度系数
FIM-01	APU 无法启动的故障隔离	ME	Ⅲ
FIM -02	飞机上故障代码为 21002000 的故障隔离	AV/ME	Ⅱ
FIM -03	飞机上故障代码为 28201101 的故障隔离	AV/ME	Ⅱ
FIM -04	飞机发动机滑油超温（琥珀色）故障隔离	ME	Ⅰ
FIM -05	飞机空调区域灯不工作的故障隔离	AV/ME	Ⅰ
FIM -06	飞机 PSEU 灯亮的故障隔离	AV/ME	Ⅰ

项目评价

"项目3　飞机防滞／自动刹车控制器故障隔离"过程考核评价标准						
课程：　　　　　　授课教师：　　　　　　专业：						
考核内容	考核标准	评价方式				
知识点（25分）： 1．故障隔离手册（FIM）的结构和内容。 2．FIM 有效性查询方法。 3．IPC 装配图和零部件相关信息查询方法。 4．FIM 隔离程序查询方法。 5．飞机防滞／自动刹车控制器相关信息查询方法。	□全对（25分）。	中国大学 MOOC 平台评分（100%）				
法规知识点（20分）： 1．飞机防滞／自动刹车控制器更换标准施工程序。 2．飞机防滞／自动刹车控制器测试标准施工程序。	□制定的施工工卡完全满足施工程序（20分）。 每错（漏）一项施工程序扣5分，扣完为止。	教师评分（100%）				
技能点（20分）： 1．飞机防滞／自动刹车控制器更换标准施工。 2．飞机防滞／自动刹车控制器测试标准施工。	□飞机防滞／自动刹车控制器更换过程规范，并完成修理任务（20分）。 每错一步施工步骤，扣5分，扣完为止。	小组自评（10%）	小组互评（20%）	教师评分（50%）	企业导师评分（20%）
工作作风　操作安全（5分）	□正确做好高空作业安全措施，安全步骤不缺失，计5分，否则计0分。				
工作作风　团结协作（5分）	□与所分配成员相处融洽，积极参与、分工协作，顺利完成任务，得5分，否则计0分。				
工作作风　节约意识（5分）	□合理使用耗材，计5分，出现浪费计0分。				
工作作风　6S 管理（5分）	□按 6S 要求整理操作台、实训场地，计5分，否则计0分。				
安全意识（5分）	□项目报告中，精准表达对安全意识的体会得5分。	教师评分（100%）				
责任使命感（5分）	□项目报告中，对责任使命感体会深刻得5分。					
职业品格（5分）	□项目报告中，对职业品格体会深刻得5分。					

项目
4

SSM 等手册应用：飞机可收放着陆灯的故障分析
Application of SSM and Other Manuals: Failure Analysis of Aircraft Retractable Landing Light

飞机的"大灯"——着陆灯

项目导读

夜晚的机场，我们总能清楚地看到机身两侧两道耀眼的白光，那就是飞机的"大灯"（着陆灯）。一般着陆灯光强可达数十万坎（1 坎等于 0.9814 烛光）。飞机着陆灯安装在飞机的两侧机翼翼根以及飞机的起落架支柱上，一般分为固定式着陆灯和收放式着陆灯两种。它们的作用和汽车的远光灯十分类似，主要用于飞机在夜间进行起飞和着陆时的前方照明，有些着陆灯还兼有滑行照明作用，可以让飞行员清晰地目测飞机和跑道的相对位置以及飞行的高度，能够更加安全地操纵飞机着陆并和地面滑行，确保飞机安全起飞和着陆。其中，收放式着陆灯功率较大，带有电动机构，具有收放功能。当飞机起飞和着陆时，收放着陆灯放出，收放着陆灯亮。当飞机在起飞和着陆时，开关打在 ON 位，但可收放着陆灯不放下，灯也不亮则表明可收放着陆灯出现故障，此时，需要遵循航空安全领域"零容忍"的原则进行维修。那么该故障的原因是什么呢？可收放着陆灯需要满足多大的工作电压？在飞机可收放着陆灯出现故障时，可以借助什么手册快速、准确地排除故障？

教学目标

★掌握系统原理图手册（SSM）的功能。
★掌握系统原理图手册（SSM）的结构。
★掌握系统原理图手册（SSM）的查询方法。
★掌握故障维修方案的制定。

学习导航

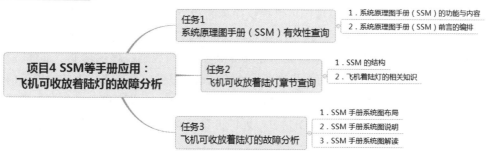

任务1　系统原理图手册（SSM）有效性查询

任务描述

飞机可收放着陆灯出现故障，需要通过系统原理图手册查询飞机灯光系统中可收放着陆灯的供电、与系统的逻辑关系、电路操作和电路工作原理等信息，以快速、准确地排除故障。系统原理图手册（System Schematic Manual，SSM）是用于展示飞机所有系统原理图示，以便理解系统原理和排除系统故障的一本手册。系统原理图手册（SSM）与飞机维修手册（AMM）和零部件图解目录（IPC）相似，也是客户化手册，按照《ATA 100 规范》编排。

如何查询飞机的有效性？SSM 与 AMM、IPC、FIM 的飞机有效性查询有何不同？

任务要求

- 了解系统原理图手册中的飞机有效性查询。
- 了解系统原理图手册中有效性的相关信息。

知识链接

SSM 能干啥？

1. 系统原理图手册（SSM）的功能与内容

系统原理图手册是由波音公司遵照《ATA 2200 规范》（基于《ATA 100 规范》）编写的客户化手册。SSM 用来辅助理解系统的功能，便于故障隔离到航线可更换件（LRU）。它并不能代替其他维护手册（如故障隔离手册 FIM、飞机维护手册 AMM、线路图手册 WDM）。

SSM 手册中，同一内容的系统图，按照机队的不同构型，会有 page101、page102、page103 等不同的页码号，查询时请选定该飞机的有效性。此外，同一回路的系统图，如果不能显示在同一页内，则会增加为 sheet1、sheet2 等，图章节号和页码号不变。

SSM 手册系统图分成三类：

（1）框图。框图提供一个系统的总概，或者系统的一部分。用于展现主要的功能和部件，组合功能和相关接口。

（2）简单原理图。简单原理图可以提供简单的功能、部件和接口，比第一类框图呈现更多的细节、功能，但并不一定包括在飞机上的位置或者销钉到销钉的线路。

（3）原理图。原理图提供系统显示足够深到 LRU 的级别，提供功能、部件、销钉到销钉的连通性和接口，提供功能和物理特性的关系，提供部件在飞机上的位置等信息。

因此，SSM 手册系统图，可提供系统所有的 LRU、LRU 位置、简单功能和内部回路，识别所有和 LRU 交叉引用的系统图索引，提供主要功能的信号传递线路或指示。

2. 系统原理图手册（SSM）前言的编排

SSM 的前言包含飞机有效性对照表（Effective Aircraft）、手册发送说明（Transmittal Letter）、修订重点说明（Highlights）、有效页清单（Effective Pages）、章节有效性（Effective Chapters）、波音修订记录（Boeing Revision Record）、修订记录（Revision Record）、临时修订记录（Record of Temporary Revision）、维修服务通告清单（Service Bulletin List）、客户更改清单（Customer Change List）、字母索引（Alphabetical Index）、简介（INTRODUCTION）等。其中简介中包括一般信息（GENERAL INFORMATION）、定义（DEFINITIONS）、系统原理（SYSTEM SCHEMATICS）。SSM 前言编排如图 4-1 所示。

考虑系统原理图手册（SSM）的前言结构与前面介绍过的 AMM、IPC、FIM 有相类似的部分，此处将不再赘述。本任务重点介绍与以往所学内容不同的部分。

（1）一般信息（GENERAL INFORMATION）：手册的适用性、介绍页的目的、系统原理图手册的作用等一般描述，如图 4-2 所示。

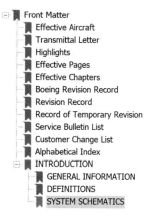

图 4-1　SSM 前言编排

737-700 SYSTEM SCHEMATIC MANUAL

INTRODUCTION

1. APPLICABILITY

This System Schematic Manual is applicable only to those Boeing airplanes listed on the Effective Aircraft page. The instructions and information contained herein apply solely to those airplanes and are not suitable for use with any other Boeing airplane(s).

2. GENERAL DESCRIPTION

This System Schematic Manual (SSM) is a collection of diagrams which define the airplane systems. These data are prepared essentially in accordance with ATA Specification No. 2200, Revision 2001.1.

This manual may also contain data and information provided by the customer. The Boeing Company assumes no responsibility for the accuracy and validity of data and information provided by a customer.

"THE BOEING COMPANY HEREBY EXPRESSLY DISCLAIMS ANY AND ALL WARRANTIES, EXPRESS OR IMPLIED, ORAL OR WRITTEN, ARISING BY LAW, COURSE OF DEALING, OR OTHERWISE, AND WITHOUT LIMITATION ALL WARRANTIES AS TO QUALITY, OPERATION, MERCHANTABILITY, FITNESS, FOR ANY INTENDED PURPOSE, AND ALL OTHER CHARACTERISTICS WHATSOEVER, OF CUSTOMER ORIGINATED MATERIAL INCORPORATED INTO THE MANUAL. THE FOREGOING DISCLAIMER SHALL ALSO APPLY TO ANY OTHER PORTION OF THIS MANUAL WHICH MAY BE AFFECTED OR COMPROMISED BY SUCH CUSTOMER ORIGINATED CHANGES."

Every effort has been made to ensure that the information presented on these schematics is complete and correct. However, in the event of conflict between this manual and Boeing Wiring Diagrams or other engineering drawings, the wiring diagrams or drawings shall be the controlling definition.

A. Purpose of Introduction Section

This Introduction Section is intended to provide the user with an overview of the SSM, an explanation of symbols used, and assumptions made while developing these schematics. Without an understanding of these symbols and assumptions, the user may not get the full value from the enclosed schematics.

B. Purpose of System Schematic Manual

The System Schematic Manual (SSM) was prepared to serve as a source of information to assist in understanding system function and to facilitate fault isolation to the Line Replaceable Unit (LRU) level. It is not intended for use as a substitute for other maintenance documentation (i.e., Fault Isolation Manual, Maintenance Manual, Wiring Diagram Manual). The SSM does not include information for testing. The procedures in the Fault Isolation Manual should be used for any fault isolation requiring testing. The procedures in the Maintenance Manual should be used to support removal and installation of components. The Wiring Diagram Manual (WDM) should be used as a reference to isolate faults in wiring and in-line disconnects.

GENERAL INFORMATION

图 4-2　一般信息简介

（2）定义（DEFINITIONS）：给出手册中使用的缩写和首字母缩写的列表，如图 4-3 所示。

 BOEING **737-700 SYSTEM SCHEMATIC MANUAL**

INTRODUCTION

The following is a list of abbreviations and acronyms used in this manual. Where marked with an asterisk (*), see the GENERAL INFORMATION section, in the Wiring Diagram manual, for additional definition information.

A/C	Air Conditioning
A/C	Aircraft
A/R	Altitude Rate
ACARS	ARINC Communications Addressing and Reporting System
ACE	Actuator Control Electronics
ACESS	Advance Cabin Entertainment and Service System
ACM	Air Cycle Machine
ACMP	Alternating Current Motor Pump (See also EMP)
ACMS	Airplane Conditioning Monitoring System
ACP	Audio Control Panel
ADF	Automatic Direction Finder
ADI	Attitude Director Indicator
ADIRS	Air Data Inertial Reference System
ADIRU	Air Data Inertial Reference Unit
ADL	Airborne Data Loader
ADM	Air Data Module
ADP	Air Driven Pump
ADRS	Address
ADS	Air Data Systems
ADU	Air Drive Unit
AEM	Audio Entertainment Multiplexer
AFDC	Air Flight Data Control

AFDS	Autopilot Flight Director System
AFL	Air Flow
AIDS	Airborne Integrated Data System
AIMS	Airplane Information Management System
AMU	Audio Management Unit
ANCMT	Announcement
ANCPT	Anticipate
ANCPTR	Anticipator
ANS	Ambient Noise Sensor
ANTI-COLL	Anti-Collision
AOA	Angle of Attack
AOC	Air/Oil Cooler
APB	Auxiliary Power Breaker
APID	Airplane Identification
APU	Auxiliary Power Unit
ARINC	Aeronautical Radio Incorporated
ASA	Autoland Status Annunciator
ASCPC	Air Supply Cabin Pressure Controller
ASCTS	Air Supply Control and Test System
ASCTU	Air Supply Control and Test Unit
ASP	Audio Select Panel
AVM	Airborne Vibration Monitor
BDY BLK	Burndy Block
BFE	Buyer Furnished Equipment
BPCU	Bus Power Control Unit
BSCU	Brake System Control Unit

DEFINITIONS

图 4-3　缩写定义简介

（3）系统原理（SYSTEM SCHEMATICS）：介绍系统原理图等级及相关内容，如图 4-4 所示。

 BOEING **737-700 SYSTEM SCHEMATIC MANUAL**

INTRODUCTION

1. LEVELS OF SCHEMATICS

Three levels of schematics may be drawn to represent the system functions:

Level 1　BLOCK DIAGRAM: Provides a broad overview of the system, or part of a system, showing major functions and components, functional groupings and pertinent interfaces.

Level 2　SIMPLIFIED SCHEMATIC: Provides a simplified view of the functions, components and interfaces. Broader in scope, showing more detail than level 1 schematics. Functions are shown without regard to their location in the aircraft or to pin-to-pin circuits.

Level 3　SCHEMATIC: Shows the system in sufficient depth for fault isolation to the LRU level. Provides a detailed view of the functions, components, pin-to-pin connectivity and interfaces. Provides a link between the function and the physical implementation. Provides the location reference for the components in the airplane.

2. CONTENT OF SCHEMATICS

The schematics show each system in a functionally integrated presentation that:

- Identifies and locates all LRU's and shows their functional internal circuitry in a simplified manner.
- Identifies connections between LRU's with cross reference to all interfacing system schematics.

- Provides signal flow for primary functions which require airplane wiring or observable indications.

The preferred schematic layout is power on the left and load on the right; signal source on the left, and signal destination/indication on the right. After satisfying proper left to right flow, the equipment is shown in relation to its position in the airplane, when possible. Left is forward, right is aft, top is right, bottom is left.

Unless otherwise noted, all schematics are shown with the airplane on the ground, after a normal flight, and with the post-flight checklist completed (power off). Instruments, indicators and monitors may reflect other conditions where clarity of presentation is improved.

Schematics may contain information relating to the nominal actuating pressure, temperature, or quantity values of certain devices, as well as dimensional relationships and operational notes. Such information is provided for reference only as an aid in systems understanding and is not intended for use to do rigging, calibration, adjustment, or functional testing. Refer to the Maintenance Manuals for this data.

A. Schematic Organization/Numbering System

ATA Specification 2200 assigns chapters to each major system (e.g., Hydraulics) of functional group of systems (e.g., Navigation). Each chapter is assigned a two-digit number (e.g., Hydraulics is Chapter 29 and Navigation is Chapter 34).

SYSTEM SCHEMATICS

图 4-4　系统原理图简介

任务实施

1. 飞机有效性查询

打开前言（Front Matter），单击飞机有效性列表（Effective Aircraft）查询飞机有效性代码（图 4-5）。

BOEING
737-700 SYSTEM SCHEMATIC MANUAL

This manual is applicable to the aircraft on this list:

| Model-Series | Operator | | Manufacturer | | | Registration Number |
	Identification Code	Effectivity Code	Block Number	Serial Number	Line Number	
737-700	XXX	001	YA811	29912	140	B-2639
737-700	XXX	002	YA812	29913	148	B-2640
737-700	XXX	003	YA813	30074	292	B-2503
737-700	XXX	004	YA814	30075	311	B 2502

EFFECTIVE AIRCRAFT

图 4-5 SSM 飞机有效性

2. 读懂线路图手册的飞机有效性列表

与项目一中提到的波音飞机维护手册类似，由图 4-5 我们可以轻松判断，飞机 B-2639 的有效性代码为 001，造册号为 YA811。

思考题

1. 若飞机的有效性列表中没有的飞机注册号，能使用该手册查询吗？
2. 你能说出 AMM、IPC、FIM、SSM 前言部分中它们的异同点吗？

任务 2 飞机可收放着陆灯章节查询

任务描述

系统原理图手册的有效性查询确认了飞机的有效性和适用性。接下来，我们将使用满足飞机有效性的 SSM 手册查询并明确飞机可收放着陆灯所在章节。

系统原理图手册（SSM）与零部件图解目录手册（IPC）、飞机维护手册（AMM）、故障隔离手册（FIM）都属于客户化手册，章节划分按照《ATA 100 规范》编排。我们只要充分熟悉《ATA 100 规范》章节划分的原则，就能快速地查询并确定飞机可收放着陆灯所在章。

你还记得《ATA 100 规范》中，飞机可收放着陆灯归属哪个系统吗？借助已学过的 AMM PART I SDS（Systems Description Section），你是否能获取飞机可收放着陆灯的相关信息？

任务要求

● 了解系统原理图手册（SSM）的结构。
● 了解飞机可收放着陆灯相关信息的查询。

知识链接

1. SSM 的结构

SSM 手册的结构和 AMM、IPC、FIMM 基本相同，手册包含前言和各系统章节原理图。前言（Front Matter）除包含了本册的有效页清单、修订改版说明、服务通告（SB），还包括了系统原理图手册一般信息简介、定义、系统原理简介等，该内容在本项目任务 1 中已详细附图介绍，该处不再赘述。SSM 手册除 21-80 系统章以外，还包含 00 章。

（1）SSM 手册 00 章。SSM 手册 00 章放置在手册所有系统图的前面，主要提供了以下三部分内容。

第一部分，00-00-00 图形符号说明。提供在 SSM 手册系统图中主要的图形符号所代表的含义以及功能，分机械和电气两类。机械类包括传感器、阀门、过滤器等，电气类包括放大器、逻辑门电路、电气开关等。图 4-6 所示为电子电气类符号。

第二部分，00-06 飞机尺寸和主要设备位置。提供飞机机身、安定面、大翼以及吊架的详细站位和部分尺寸。另外，提供飞机上主要设备中心的位置，如图 4-7 所示。

第三部分，地面勤务接近盖板。为了便于日常的勤务工作，这些盖板分布在飞机各处，方便接近相应的设备，如电子舱门、水勤务面板、大翼各注油（润滑）点、地面空调引气接口等。

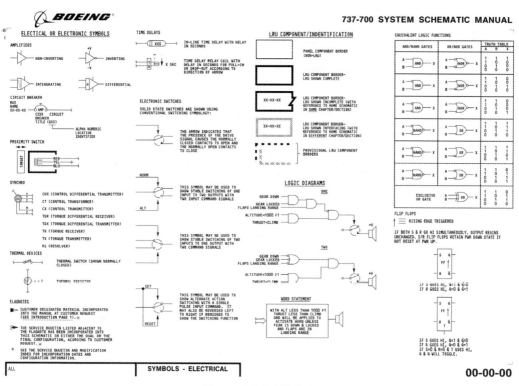

图 4-6　图形符号说明

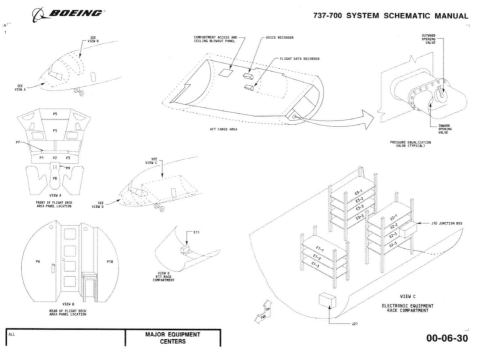

图 4-7　设备架位置图

（2）SSM 手册系统章。SSM 手册各系统章包括两部分，前部内容和系统章节原理图。如图 4-8 所示，前部内容包含有效页清单、目录、字母索引等。

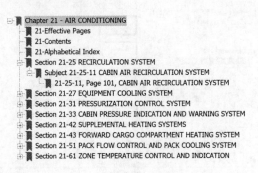

图 4-8　SSM 手册系统章

如图 4-9 所示，系统章目录中给出了该章中所有原理图的条目，可以方便使用者快速查询。

BOEING　　　　　　　　　　　　　**737-700 SYSTEM SCHEMATIC MANUAL**

CHAPTER 21
AIR CONDITIONING

Title	CH-SC-SU	Schem	Page	Sheet	Date	Effectivity
RECIRCULATION SYSTEM						
CABIN AIR RECIRCULATION SYSTEM	21-25-11		101		Jan 13/2006	ALL
EQUIPMENT COOLING SYSTEM						
EQUIPMENT COOLING - EXHAUST	21-27-21		101	1	Jan 13/2006	ALL
				2	Jan 13/2006	ALL
EQUIPMENT COOLING - SUPPLY	21-27-31		101	1	Jan 13/2006	ALL
				2	Jan 13/2006	ALL
PRESSURIZATION CONTROL SYSTEM						
PRESSURIZATION CONTROL MANUAL MODE	21-31-11		101		Jan 13/2006	YA811-YA814
PRESSURIZATION CONTROL AUTO 1	21-31-22		101	1	Jan 13/2006	YA811-YA814
				2	Jan 13/2006	YA811-YA814
PRESSURIZATION CONTROL AUTO 2	21-31-23		101	1	Jan 13/2006	YA811-YA814
				2	Jan 13/2006	YA811-YA814
PRESSURIZATION CONTROL AUTO CHANNEL INTERFACES	21-31-24		101		Jan 13/2006	YA811-YA814
PRESSURIZATION CONTROL LCD LIGHTING	21-31-25		101		Jan 13/2006	ALL
CABIN PRESSURE INDICATION AND WARNING SYSTEM						
CABIN PRESSURE WARNING	21-33-11		101		Jan 13/2006	ALL
SUPPLEMENTAL HEATING SYSTEMS						
DOOR AREA HEATERS	21-42-11		101		Jan 13/2006	YA811-YA814
FORWARD CARGO COMPARTMENT HEATING SYSTEM						
OVERBOARD EXHAUST VALVE CONTROL	21-43-21		101		Jan 13/2006	ALL

21-CONTENTS

图 4-9　SSM 手册系统章目录

2. 飞机着陆灯的相关知识

飞机在夜航进近中，需要借助灯光照明与指示确保安全着陆。尽管大量的研究表明，飞行员可以在不需要着陆灯的情况下，仅通过机场助航灯光系统（地面灯光）的照明和指示，进行跑道位置判断识别，

教你通过着陆灯的位置辨别波音、空客、C919

并安全完成夜航着陆。但是，助航灯光更多只能起到基本的指示作用，并不能让飞行员更清楚地了解跑道道面及周边环境。根据 SAE-ARP693（《着陆和滑行灯安装设计标准》）

提出的飞机着陆灯作用说明，着陆灯为飞机进近、触地、滑行、起飞提供环境参考信息，在飞机夜航进近滑行时照射机场跑道上的主要障碍物，因此机场助航灯光系统必须有着陆灯配合使用才能更好地遵循航空安全领域"零容忍"的最高原则。

着陆灯按结构形式可分为固定式和收放式两种。波音 737 的收放式着陆灯位于机腹靠近翼根的位置，固定式着陆灯安装于机翼前缘。而空客 A320 因其机翼前缘有缝翼，无法安装固定式着陆灯，只能在机翼翼根下部安装可收放式着陆灯。一般可收放式的着陆灯都安装于有开口处的表面，以方便在收起位置时能够收缩在机身轮廓之内，不影响气动布局。有的飞机还把固定式着陆灯安装于起落架构件上，用于着陆后对跑道道面提供大面积照明。我国 C919 飞机的着陆灯与波音 737 一样，采用气动融合式设计，位于机翼翼根处，并在此基础上有一定的自主创新。部分小型单发螺旋桨飞机的着陆灯就安装在机鼻处，简单且高效。

细心的你可能已经发现，在天气晴朗的白天，飞机也会开启着陆灯起降。这是因为，根据中国民航总局规定，各航空公司机组在 3000 米（约 10000 英尺）以下高度飞行需要打开着陆灯，其主要目的是防止鸟击。有调查显示，每年全世界大约会发生 2 万起飞机鸟击事故，全球民航业每年因鸟击而造成的直接经济损失约为 12.8 亿美元，鸟击已经成为航空业三大自然灾害之一，因此避免鸟击被视为飞行安全领域的重中之重。与噪声一样，灯光也有一定的驱鸟作用，因此，尽管在着陆灯设计指标中没有关于驱鸟效果的明文规定，但中国民航仍然规定飞机在 3000 米以下高度飞行打开着陆灯。但是，并非所有鸟类都会在灯光的刺激下产生本能的躲避行为。美国内布拉斯加州林肯大学的研究人员对脉冲式白光着陆灯在刺激鸟类躲避行为的课题进行了研究，发现仅有部分鸟类对"脉冲光"会产生本能的躲避行为。为了减小鸟击事故的发生概率，科学家正致力于对包含特殊波长和脉冲频率的着陆灯进行研究，这种灯光可用于刺激鸟类的视觉感官并使其产生本能的躲避行为，相信不久的将来，飞机的着陆灯在保障照明的同时，也可以有针对性地起到防止鸟击的作用。

任务实施

1. 飞机可收放着陆灯相关信息查询

对于飞机可收放着陆灯相关信息，可查询满足飞机有效性的 AMM PART I SDS（Systems Description Section）33 章灯光系统（LIGHTS）- 着陆灯（LANDING LIGHTS）- 可收放着陆灯（RETRACTABLE LANDING LIGHTS）。如图 4-10 所示，可收放着陆灯的作用是让飞行员在起飞和降落时看清跑道，其控制开关在头顶面板 P5。

打开可收放着陆灯（Retractable Landing Lights）和固定着陆灯（Fix Landing Lights）功能说明，可找到飞机可收放着陆灯的相关信息。如图 4-11 所示，三挡拨动开关控制可收放着陆灯。可收放着陆灯的工作电压为 115V 交流，当可收放着陆灯开关在 EXTEND位置时，115V 交流带动电机，当开关在 ON 位，可收放着陆灯放下，灯向外延伸超过5°，可收放着陆灯就会点亮。

BOEING   737-600/700/800/900 AIRCRAFT MAINTENANCE MANUAL

LIGHTS - LANDING LIGHTS - RETRACTABLE LANDING LIGHTS

Purpose

The landing lights help the pilots see the runway during takeoff and landing.

Physical Description

The retractable landing light has these parts:

- Lens assembly
- Lamp
- Retainer screws
- Extend/retract motor.

Location

The retractable landing lights are on the fuselage, adjacent to the ram air inlet panels.

The control switches for the retractable landing lights are on the P5 forward overhead panel.

Training Information Point

You must extend the retractable landing light to replace the lamp. After you extend the retractable landing light, pull and collar the circuit breaker to prevent accidental operation of retraction mechanism or electric shock.

EFFECTIVITY
XXX ALL

33-42-00

图 4-10　可收放着陆灯相关信息

BOEING   737-600/700/800/900 AIRCRAFT MAINTENANCE MANUAL

LIGHTS - LANDING LIGHTS - RETRACTABLE AND FIXED LANDING LIGHTS FUNCTIONAL DESCRIPTION

Operation

A two-position toggle switch controls the fixed landing lights.

A three-position toggle switch controls the retractable landing lights. These are the positions of the switch:

- RETRACT; the light retracts and is off
- EXTEND; the light extends and the light is parallel to the aircraft water line
- ON; the light is on.

The light will come on only with the switch in the ON position.

Functional Description

When the fixed landing light switch is in the ON position, 115v ac goes to the step down transformer. The step down transformer decreases the voltage to 28v ac.

The retractable landing light uses 115v ac to extend and retract the light.

When the retractable landing light switch is in the RETRACT position, 115v ac goes to the retract motor in the light. The light retracts until the full retract limit switch opens.

When the switch is in the EXTEND position, 115v ac goes to the extend motor in the light. The light extends until the full extend limit switch opens.

When the switch is in the ON position, the landing light extends. When the light is within five degrees of full extension the light will come on.

Training Information Point

WARNING: MAKE SURE YOU SEAL THE TERMINALS TO PREVENT AN EXPLOSION OF THE FUEL FUMES. AN EXPLOSION CAN CAUSE INJURY TO PERSONS AND DAMAGE TO EQUIPMENT.

CAUTION: LANDING LIGHTS ARE NOT DESIGNED FOR CONTINUOUS USE IN STILL AIR. LIMIT OPERATION TO MOMENTARY USE.

EFFECTIVITY
XXX ALL

33-42-00

图 4-11　可收放着陆灯功能说明

2. 飞机可收放着陆灯章节查询

打开 SSM 手册 33 章灯光系统的目录（Contents），可查询飞机可收放着陆灯所在章节，如图 4-12 所示。

飞机可收放着陆灯章节查询

BOEING   **737-700 SYSTEM SCHEMATIC MANUAL**

CHAPTER 33
LIGHTS

Title	CH-SC-SU	Schem	Page	Sheet	Date	Effectivity
PASSENGER AND LAVATORY CALL - RIGHT FWD	33-27-31		101		Jan 13/2006	ALL
PASSENGER AND LAVATORY CALL - RIGHT AFT	33-27-32		101		Jan 13/2006	ALL
PASSENGER AND LAVATORY CALL - PSU	33-27-41		101		Jan 13/2006	ALL
ENTRY LIGHTS						
ENTRY LIGHTS	33-29-11		101		Jan 13/2006	ALL
WHEEL WELL LIGHTS						
SERVICE LIGHTING - WHEEL WELLS	33-32-11		101		Jan 13/2006	ALL
AIR CONDITIONING COMPARTMENT LIGHTS						
SERVICE LIGHTING - AIR CONDITIONING COMPARTMENT	33-33-11		101		Jan 13/2006	ALL
ELECTRONIC EQUIPMENT COMPARTMENT LIGHTS						
SERVICE LIGHTING - EQUIPMENT RACK FORWARD LOWER COMPARTMENT	33-34-11		101		Jan 13/2006	ALL
ACCESSORY COMPARTMENT LIGHTS						
SERVICE LIGHTING - ACCESSORY COMPARTMENT	33-35-11		101		Jan 13/2006	YA811-YA814
CARGO COMPARTMENT LIGHTS						
SERVICE LIGHTING - CARGO COMPARTMENT - FWD	33-36-11		101		Jan 13/2006	ALL
SERVICE LIGHTING - CARGO COMPARTMENT - AFT	33-36-12		101		Jan 13/2006	ALL
WING ILLUMINATION LIGHTS						
EXTERIOR LIGHTS - WING SCANNING	33-41-11		101		Jan 13/2006	ALL
LANDING LIGHTS						
EXTERIOR LIGHTS - LANDING	33-42-11		101		Jan 13/2006	ALL

33-CONTENTS

图 4-12　SSM 33 章目录

因此，可确认飞机可收放着陆灯系统原理图在 33-42-11。

思考题

1. 如果标志灯（Logo Lights）故障，你知道应该如何确定章节吗？
2. 如果陌生的设备出现故障，你知道如何在手册中获得帮助吗？

任务 3　飞机可收放着陆灯的故障分析

任务描述

已知飞机在起飞和着陆时，可收放着陆灯出现故障，开关打在 ON 位，但可收放着陆灯不放下，灯也不亮。该故障需要借助系统原理图手册（SSM）分析可收放着陆灯电路原理，进行故障维修。上一任务中，我们已经知晓了飞机可收放着陆灯系统原理图所在的章节。为了排除可收放着陆灯故障，我们还需要深入探寻飞机可收放着陆灯系统原理图，分析系统线路及线路中设备的连接情况。

如何利用 SSM 手册查询故障所在系统原理图？如何分析 SSM 手册中的原理图？

任务要求

- 掌握系统原理图的查询。
- 掌握系统原理图的分析与排故。

知识链接

1. SSM 手册系统图布局

SSM 手册系统图的布局通常是：电源在左侧，负载在右侧；信号源在左侧，信号接收或显示在右侧。为满足基本的从左到右的布局，系统图上显示设备在飞机上的位置，通常左为前侧，右为后侧，上为右侧，下为左侧，例如 LOGO 灯系统图，如图 4-13 所示。

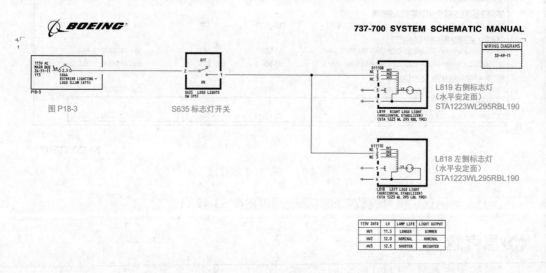

图 4-13　LOGO 灯系统图

除非特殊说明，所有的系统图均为飞机在地面的状态，正常飞行后，完成飞行后检查，系统断电。仪表、显示和监控可能在其他所需要呈现的状态。

SSM 手册系统图可能会包含其他的相关信息，例如启动压力、温度或者具体设备的量值，此外还有其他相关信息和操作提示。这些都是用来增加对系统的理解但仅供参考，不能以此作为装配、校准、调节和功能测试的依据，这些数据请参考对应的维护手册（AMM）。

如图 4-13 所示，LOGO 灯系统图的每一个设备中，除了标注设备名称外，在每一个部件下方左侧均有一个由字母和数字组成的编号，如 S635- 标志灯开关、L818- 左侧标志灯、L819- 右侧标志灯。这些编号代表具体的部件，此编号就是设备号。设备号以字母开头，由字母和数字共同组成，最多包含 10 个字母和数字，字母前面可能会有空格（9000-9999 和 90000-99999 的设备号是给客户预留的）。设备类型与代号的对应关系见表 4-1。

表 4-1　设备类型与代号的对应关系

设备号	设备类型
A	防冰设备、除冰轮胎罩、防冰控制开关、皮托管加热器、防冰探测器、NESA 窗
B	黑匣子
C	电路开关、其他保护装置
D	插头
E	设备架
F	燃油系统
G	发电机
GD	接地 - 机身（GD）
J	接线盒
L	各种灯照明装置及灯组件
M	附件组件、天线及天线调节器、放大器及电池、镇流器组件、铃及谐音、计算机、电容、控制组件、方向陀螺、滤波器、磁通活门、扬声器、ILS 设备架、卫生间组件、马达、泵、相位适配器、电源组件、接收机、寻呼及寻呼协议、磁带录音机、发送机、垂直陀螺、水加热器、其他设备
N	指示器
P	面板
R	继电器、变阻器、电气电阻、整流器、电位差计
S	开关
SM	拼接管（同一线束内）
SP	拼接管（线束之间）
T	变压器、发送器、温度灯泡、接线条
TB	接线块
V	阀门
Y	航线可更换件

注意：任何设备均不用 W 做设备号。

设备号分配给每一个和线路相接的 LRU、面板和设备架。并不是所有具有设备号的部件都是 LRU，也不是所有的 LRU 都有设备号。设备号代表该部件的功能、位置等所有特性，设备号用来标识部件。同时有专门的设备清单 EQ（Equipment List）以设备号排序，来汇总飞机上设备的件号、名称、位置、所在章节等信息。设备号在飞机所有的手册中通用，设备清单在 WDM（Wiring Diagram Manual）手册中查询，此内容在项目 5 中有详细介绍。

2. SSM 手册系统图说明

SSM 手册系统图中涉及线路说明、插头说明、定位、数据总线等信息。

（1）线路说明。系统图上所示意的连接在设备之间的线路，并不显示系统所有相关的回路或显示线束中所有的线。对于本图未显示的回路，用章节号索引到该回路所在的系统图中。插头中其他插钉所连接的同一章节号用大括号来表示，并用大括号来表示所包含的线束。如图 4-14 所示，D1679 的 Pin1、Pin2、Pin3、Pin4、Pin5 原理图在 22-12-41。

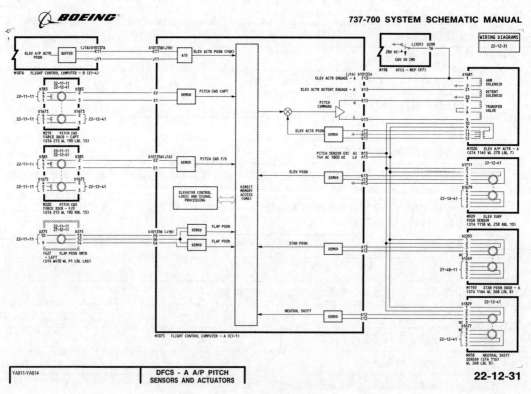

图 4-14　系统图示例

（2）插头说明。线路通过插头连接到航线可更换件 LRU，具体的线路对应着插头上具体的销钉。如图 4-13 中所示，插头的设备号通常以 D 开头，在插头销钉号的上方显示。

插头固定在设备架上，通常 LRU 的更换不会影响到销钉或线路，只需要对接好插头即可。系统图中标注出插头内部销钉的连接，以便于系统的分析和排故。

（3）定位。每一个设备的安装位置通过插图说明，或者通过设备描述说明。位置可能是面板或设备架、通用的飞机区域划分、舱门的定位或机身站位。如图 4-14 所示，飞行控制计算机 B（Flight Control Computer B）在设备架 E1-4。

（4）数据总线。系统图上的总线用一组平行线表示，用箭头表示数据流的方向，平行线表示 LRUS 之间的数据总线连接。

3. SSM 手册系统图解读

分析和解读 SSM 手册系统图，需掌握系统图中跳开关、二极管、延时控制器、延时继电器、压力电门、逻辑门等详细信息。

（1）跳开关。跳开关相当于保险，如图 4-15 所示，28V AC BUS 1 表示此跳开关通过 1 号 28V 交流汇流条供电。7.5 表示额定跳开电流，即保护电流为 7.5A；C331 是跳开关的设备号，在 WDM 中可以利用设备号查询该跳开关的件号；跳开关下面的文字描述是这个跳开关控制的系统；P6-3 和 B9 是跳开关的位置在面板 P6-3 的 B 行 9 列。

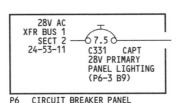

图 4-15　跳开关

利用 SSM 排故很重要的步骤就是，了解各个系统的汇流条供电信息。

（2）二极管。二极管的作用是单向导电性，如图 4-16 所示，电流只能从左向右通过，而不能从右向左。有点二极管还起到分压的作用，相当于一个小电阻。

（3）延时控制器。延时控制器如图 4-17 所示，当右端出现高电平后，2 秒后左端出现高电平。

图 4-16　二极管　　　　　　　　　　　　　　图 4-17　延时控制器

（4）延时继电器。延时继电器如图 4-18 所示，当 X1 得电后，10 秒后继电器得电吸合。

（5）压力电门。压力电门如图 4-19 所示，当压力小于 750PSI 时，插头 D11730 的 3 号钉通过电门与 2 号钉连接，即 3 号钉接地；当压力电门感受压力大于 750PSI 时，电门朝箭头所指方向运动，此时 1 号钉与 2 号钉连接接地。由于是三角形触点，因此当压力恢复小于 750PSI 时，电门回到原来位置，保持在圆形触点上。

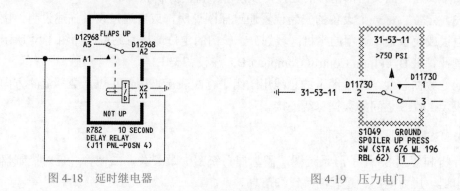

图 4-18　延时继电器　　　　　　　　　　　图 4-19　压力电门

（6）逻辑门。着陆警告逻辑图如图 4-20 所示，两幅图的输入信号一致，输出信号也一致，只是中间的逻辑控制不同。它们通过不同的逻辑门来控制，得到相同的控制结果。

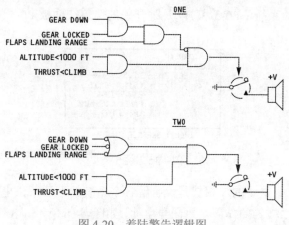

图 4-20　着陆警告逻辑图

　　第一幅图左上角是与门，只有当输入信号同时为 1（高电位）时，输出才是 1，所以当起落架放下并锁好时输出 1，此时与下面一个条件襟翼是否在着陆角度一起作为输入信号给下一个与门。当襟翼在着陆角度时，下一个与门输出 1，经过一个非门后，1变成 0，此时与门下面一个与门给出的信号（高度小于 1000FT，推力小于爬升推力）同时送给下一个与门，来控制着陆警告的产生。如果上面一路中三个信号有一个信号为 0，则最后一个与门输出为 1，触发着陆警告。第二幅图原理也是一样的，只是使用了或非门来控制。

任务实施

1. 飞机着陆灯原理图分析

本项目任务 2 已确定飞机可收放着陆灯电路原理图在 33-42-11 章，因此，我们可以打开 SSM 33-42-11 EXTERIOR LIGHTS-LANGDING。

如图 4-21 所示，波音 737 飞机外部着陆灯包括固定着陆灯（FIXD LANDING LIGHTS）和可收放着陆灯（RETRACTABLE LANGDING LIGHTS）。同时，固定着陆灯与可收放着陆灯都是左右对称电路。

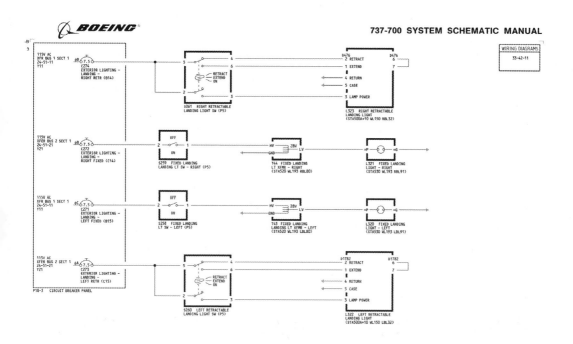

图 4-21 外部着陆灯原理图

2. 飞机可收放着陆灯故障分析

因该项目左右可收放着陆灯故障为对称图形，我们可以任意截取图 4-21 的一侧进行分析。如图 4-22 所示，为飞机右侧可收放着陆灯电路图。

飞机可收放着陆灯故障分析

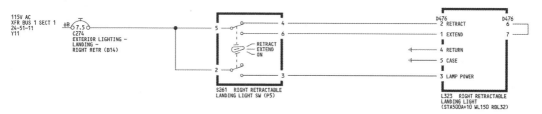

图 4-22 右侧可收放着陆灯原理图

通过分析可知，右侧可收放着陆灯由 1 号 115 交流汇流条供电，跳开关设备号为 C274，额定电流为 7.5A。当右侧可收放着陆灯开关打在 ON 位时，S261 的 2 号与 3 号接通，如果可收放着陆灯不放下，灯也不亮，有五种可能性：① S261 的 5 号与 6 号未接通；② S261 的 6 号与 L323 的 1 号断线；③ L323 的 4 号接地装置故障；④ S261 的 3 号与 L323 的 3 号断线；⑤ L323 的 5 号接地装置故障。

思考题

1. 如果飞机再循环风扇故障，你能完成再循环风扇系统相关原理图的分析吗？
2. 如果是固定着陆灯故障，你能利用本项目的查询结果说明故障可能的原因吗？

工卡实操

本项目配套民航机务维修典型案例——SSM 应用工卡 6 个，专业面向包括 AV 和 ME，难度系数设置了Ⅲ（易）、Ⅱ（中）、Ⅰ（难）三个等级，学习者可按需选用。工卡实操过程中，学习者不仅需要关注修理任务本身，同时需有意识地查询手册中相关任务的保障人身与设备安全的警告（Warning）和注意事项（Caution/Note）。工作中，工作者需要确认已经查询或知晓了相关注意事项和警告，并在工卡上签署确认，才可开始相应任务。本项目配套工卡详情见表 4-1。

表 4-1　项目 4 配套工卡列表

工卡序号	项目名称	专业方向	难度系数
SSM-01	飞机可收放着陆灯故障原理分析	AV/ME	Ⅲ
SSM -02	飞机客舱空气循环系统的原理分析	AV/ME	Ⅱ
SSM -03	飞机空中交通管制（ATC）应答机的原理分析	AV/ME	Ⅱ
SSM -04	飞机气象雷达冷却风扇的工作原理分析	AV/ME	Ⅰ
SSM -05	飞机再循环风扇工作原理分析	AV/ME	Ⅰ
SSM -06	飞机厕所烟雾探测器的工作原理分析	AV/ME	Ⅰ

➡️ 项目评价

考核内容	考核标准	评价方式			
	"项目 4　飞机可收放着陆灯的故障分析"过程考核评价标准
	课程：　　　　　授课教师：　　　　　专业：
知识点（25 分）： 1. 系统原理图手册（SSM）的结构和内容。 2. SSM 有效性查询方法。 3. IPC 装配图和零部件相关信息查询方法。 4. FIM 隔离程序查询方法。 5. AMM 施工程序查询方法。 6. 飞机可收放着陆灯相关信息查询方法。	□全对（25 分）。		中国大学 MOOC 平台评分（100%）
法规知识点（20 分）： 1. 飞机可收放着陆灯灯泡更换标准施工程序。 2. 飞机可收放着陆灯更换后测试标准施工程序。	□制定的施工工卡完全满足施工程序（20 分）。 每错（漏）一项施工程序扣 5 分，扣完为止。		教师评分（100%）
技能点（20 分）： 1. 飞机可收放着陆灯灯泡更换的标准施工。 2. 飞机可收放着陆灯测试的标准施工。	□飞机可收放着陆灯灯泡更换过程规范，并完成修理任务（20 分）。 每错一步施工步骤，扣 5 分，扣完为止。	小组自评（10%）	小组互评（20%）	教师评分（50%）	企业导师评分（20%）
工作作风　操作安全（5 分）	□正确做好高空作业安全措施，安全步骤不缺失，计 5 分，否则计 0 分。				
工作作风　团结协作（5 分）	□与所分配成员相处融洽，积极参与、分工协作，顺利完成任务，得 5 分，否则计 0 分。				
工作作风　节约意识（5 分）	□合理使用耗材，计 5 分，出现浪费计 0 分。				
工作作风　6S 管理（5 分）	□按 6S 要求整理操作台、实训场地，计 5 分，否则计 0 分。				
安全意识（5 分）	□项目报告中，精准表达对安全意识的体会得 5 分。		教师评分（100%）
责任使命感（5 分）	□项目报告中，对责任使命感体会深刻得 5 分。				
职业品格（5 分）	□项目报告中，对职业品格体会深刻得 5 分。				

随手笔记

项目
5

WDM 等手册应用：襟翼位置传感器线路故障的排除

Application of WDM and Other Manuals: Troubleshooting of Wiring Fault of Flap Position Sensor

 项目导读

　　线路故障是飞机故障中常见的一种故障，随着民用航空业的发展，老龄飞机运行过程中各类故障日益增加，其中线路故障尤为突出。如何利用手册进行故障分析并排除线路故障，是每个成熟的机务维修人应当掌握的基本维修技能。本项目根据波音飞机出现的具体故障，结合相关手册，对线路故障进行分析并排除。线路故障表现形式多种多样，如仪表显示器显示不正常、系统不工作、信号不正常、故障信息不定期出现等。维修人员能否快速而有效地对故障进行分析与排除，直接关系到航空公司飞机的利用率和经济效益。本项目结合 B737-700 型飞机相关手册与具体的线路故障，介绍线路故障分析与故障排除的过程。

　　波音 737-700 飞机的故障现象出现，自动飞行控制系统 B 中 SNSR EXC AC 跳开关跳出，不能复位，相关维护信息为 FLAP POSN-2 (J1B-G06, F06, E06)。通过 FIM 查询，故障可能的原因为设备故障、断路器故障、电源故障、线路故障。前三者故障均可使用 AMM 判断其好坏并排除。那线路故障该如何分析并排除呢？

 教学目标

　　★掌握线路图手册（WDM）的功能。
　　★掌握线路图手册（WDM）中各种清单的使用。
　　★掌握线路图手册（WDM）的查询方法。
　　★掌握线路维修方案的制定。

学习导航

项目5 WDM等手册应用：
襟翼位置传感器线路故障的排除

任务1
线路图手册（WDM）有效性查询
　　1．线路图手册（WDM）的功能与内容
　　2．线路图手册（WDM）前言的编排

任务2
襟翼位置传感器相关线路分析
　　1．WDM 的结构
　　2．襟翼位置传感器的相关知识

任务3
襟翼位置传感器线路故障的排除
　　1．WDM 手册常用清单
　　2．飞机线路故障排除的相关知识

任务 1　线路图手册（WDM）有效性查询

任务描述

飞机出现线路故障，需要通过线路图手册查询线路图纸、电气设备清单、导线清单、杂项清单、接地桩清单等，确定脱开支架插头位置等信息，以便快速、准确地排除导线短路、插钉不导通、接线片损坏等故障。线路图手册（Wiring Diagram Manual，WDM）是针对飞机各系统中线路的连接情况编写的一本手册。线路图手册（WDM）与飞机维护手册（AMM）、零部件图解目录（IPC）和故障隔离手册（FIM）相似，也是客户化手册，飞机系统线路图纸按照《ATA 100 规范》编制。

如何查询飞机的有效性？ WDM 与 AMM、IPC 和 FIM 的前言结构有何不同？

任务要求

- 了解线路图手册中飞机有效性查询。
- 了解线路图手册中有效性的相关信息。

知识链接

WDM 能干啥？

1. 线路图手册（WDM）的功能与内容

线路图手册（WDM）详细描述了飞机上的导线连接，用来排除和导线相关的故障。WDM 是由波音公司遵照《ATA 2200 规范》（基于《ATA 100 规范》）编写的客户化手册。它包含了波音飞机上的图表、图纸、相关设备导线和连接的清单。除非特殊说明，所有的线路图均为飞机在地面的状态，正常飞行后，完成飞行后检查，系统断电。

WDM 手册除了前言、00 章、21 ~ 80 系统章节外，还包括 91 章和设备清单（Equipment List，EL）。91 章提供了飞机站位布局，各面板、设备架的位置，各跳开关板图示，主要线束的布线，以及各种和导线相关的部件的清单，如跳开关清单（Circuit Breaker List）、接线管清单（Splice List）、主线束清单（Master Bundle List）、备用线清单（Spare Wire List）、脱开支架清单（Disconnect Bracket List）等。

2. 线路图手册（WDM）前言的编排

WDM 的前言包括飞机有效性列表（Effective Aircraft）、手册发送说明（Transmittal Letter）、修订重点说明（Highlights）、有效页清单（Effective Pages）、章节有效性（Effective Chapters）、波音修订记录（Boeing Revision Record）、修订记录（Revision Record）、临时修订记录（Record of Temporary Revision）、维修服务通告清单（Service Bulletin List）、客户更改列表（Customer Change List）、字母索引（Alphabetical Index）、简介（INTRODUCTION）。其中，简介里包含一般信息（GENERAL INFORMATION）、

定义（DEFINITIONS）、设备清单（EQUIPMENT LIST）、线路图（WIRING DIAGRAMS）、图表和列表（CHARTS AND LISTS）、代码（CODES）、手册的使用（MANUAL USAGE）、标准线路施工（STANDARD WIRING PRACTICE)等。WDM 前言编排如图 5-1 所示。

考虑线路图手册的前言结构与前面介绍过的 AMM、IPC、FIM 有类似的部分，此处将不再赘述。本任务重点介绍与以往所学内容不同的部分。

（1）一般信息（GENERAL INFORMATION）：介绍线路图手册的功能及编排规则，如图 5-2 所示。

（2）定义（DEFINITIONS）：给出手册中使用的缩写和首字母缩写的列表，如图 5-3 所示。

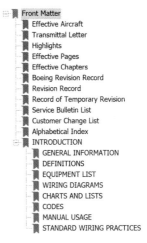

图 5-1　WDM 前言编排

737-700 WIRING DIAGRAM MANUAL

INTRODUCTION

1. APPLICABILITY

This Wiring Diagram Manual is applicable only to those Boeing airplanes listed on the Effective Aircraft page. The instructions and information contained herein apply solely to those airplanes and are not suitable for use with any other Boeing airplane(s).

2. GENERAL DESCRIPTION

The Boeing Wiring Diagram Manual (WDM) is a collection of diagrams, drawings, and Lists which define the wiring and hookup of associated equipment installed on the listed Boeing airplanes. These data are prepared essentially in accordance with the ATA Specification No. 2200, revision 2001.1.

This manual may also contain data and information provided by the customer. The Boeing Company assumes no responsibility for the accuracy and validity of data and information provided by a customer.

The WDM document number is unique to the customer whose name appears on the title page. Each chapter is preceded by its own Table of Contents (TOC), List of Effective Pages (LEP), and Alphabetical Index.

NOTE: System Schematics reside in a separate System Schematics Manual. Standard Wiring Practices–Chapter 20 reside in a separate Standard Wiring Practices manual (D6-54446).

All Wiring Diagrams are shown, unless otherwise specified, with the airplane on the ground, after normal flight, with the shutdown checklist complete (power off).

3. PROCESS CONTROLS

Control of the various manufacturing and installation processes used for wiring the airplane is covered in D6-36911 - Electrical Wiring Assembly and Installation Processes.

4. BOEING CHANGE DEFINITIONS

Changes used by Boeing to implement airplane changes that may affect this manual are listed below.

A. Customer Originated Changes (COC)

Customer Originated Changes are requests to incorporate airplane data, information, changes and modifications authorized by a customer into the WDM.

NOTE: Boeing will not undertake to test or evaluate, in any form, the validity or the technical accuracy of Customer Originated Changes. This will remain the sole responsibility of the customer submitting the Customer Originated Change request.

B. Service Bulletin (SB)

Service Bulletins provide information for accomplishing an engineering change on in-service airplanes. Service Bulletins are incorporated into this manual only upon customer request.

GENERAL INFORMATION

图 5-2　WDM 一般信息简介

737-700
WIRING DIAGRAM MANUAL

INTRODUCTION

The following is a list of abbreviations and acronyms used in this manual. Where marked with an asterisk (*), see the GENERAL INFORMATION section, in the Wiring Diagram manual, for additional definition information.

A/C	Air Conditioning
A/C	Aircraft
A/R	Altitude Rate
ACARS	ARINC Communications Addressing and Reporting System
ACE	Actuator Control Electronics
ACESS	Advance Cabin Entertainment and Service System
ACM	Air Cycle Machine
ACMP	Alternating Current Motor Pump (See also EMP)
ACMS	Airplane Conditioning Monitoring System
ACP	Audio Control Panel
ADF	Automatic Direction Finder
ADI	Attitude Director Indicator
ADIRS	Air Data Inertial Reference System
ADIRU	Air Data Inertial Reference Unit
ADL	Airborne Data Loader
ADM	Air Data Module
ADP	Air Driven Pump
ADRS	Address
ADS	Air Data Systems
ADU	Air Drive Unit
AEM	Audio Entertainment Multiplexer
AFDC	Air Flight Data Control
AFDS	Autopilot Flight Director System
AFL	Air Flow
AIDS	Airborne Integrated Data System
AIMS	Airplane Information Management System
AMU	Audio Management Unit
ANCMT	Announcement
ANCPT	Anticipate
ANCPTR	Anticipator
ANS	Ambient Noise Sensor
ANTI-COLL	Anti-Collision
AOA	Angle of Attack
AOC	Air/Oil Cooler

DEFINITIONS

图 5-3 缩写定义简介

（3）设备清单（EQUIPMENT LIST）：给出设备清单中包含的设备号、件号、供应商代码等编排规则，如图 5-4 所示。

737-700
WIRING DIAGRAM MANUAL

INTRODUCTION

1. EQUIPMENT LIST—GENERAL

Electrical and electronic equipment are shown on wiring diagrams and schematics with alphanumeric designators. These designators are used as cross-reference symbols to the Equipment List where the Part Numbers and Part Descriptions are shown. Splices, grounds, terminals and wire bundles are not included in the Equipment List.

A. Equipment List Data Fields

(a) EQUIP Field (Equipment Number)

The Equipment Number field may contain up to ten (10) alphanumeric characters. The Equipment Number always begins with a letter and may contain a space followed by another character.

NOTE: Equipment numbers 9000 through 9999 and 90000 through 99999 are reserved for customer use. Using these customer assigned equipment numbers facilitates Identification of customer installed equipment. Customers should use only customer assigned equipment numbers, not Boeing assigned equipment numbers reported in the Equipment List.

The following list shows the categories assigned to the Basic Equipment Designators.

EQUIPMENT DESIGNATOR	TYPE OF EQUIPMENT
A	Anti-Icing Controls
	Anti-Icing Equipment
	De-Icing Boots
	Ice Detector
	NESA Windows
	Pitot Heater
B	Black Box
C	Circuit Breakers
	Protective Equipment
D	Connectors
E	Equipment Racks
F	Fuel system components
G	Generator
GD	Grounds (Airframe)
H	Overflow Categories, Miscellaneous Bundle Equipment
J	Junction boxes
L	Lamp Assemblies
	Lights, Lamps

EQUIPMENT LIST

图 5-4　设备清单简介

（4）线路图（WIRING DIAGRAMS）：介绍线路图的基本信息，包括线路图编号、图表编号等，如图 5-5 所示。

737-700
WIRING DIAGRAM MANUAL

INTRODUCTION

1. **BASIC INFORMATION ABOUT WIRING DIAGRAMS**

   A. **Wiring Diagram And Page Numbering**

   (a) **Wiring Diagram Numbering**

   The Diagram numbering is in accordance with ATA Specification 2200 Revision 2001.1

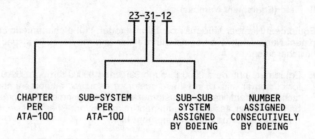

The first three digits will be identical on diagrams and schematics.

NOTE: When a diagram is referenced to another, only the diagram number is used. Therefore, where there is more than one page of the same diagram, it is necessary to refer to the effectivity block to make certain the diagram applies to the airplane of interest.

   (b) **Diagram Page Numbering**

   Diagram page numbering begins at 1 then 2, 3 etc. Each page reflects different delivered configurations between aircraft. See the following example.

DIAGRAM	PAGE	EFFECTIVITY
21-31-12	1	001-004
21-31-12	2	005-999

   The Page numbers (Page 101, 102, etc.) are used to represent different delivered configurations of a given schematic which may be applicable to different airplanes within the customer's fleet. When a schematic page number has a suffix (e.g., 101A, 102A for Customer Originated Changes or 101.1, 102.1, etc. for Service Bulletins) it reflects a post-delivery configuration for the same airplane(s). Both the configuration delivered by Boeing and the configuration after modification remain in the manual until the airline notifies Boeing that the post-delivery change has been incorporated in the customer's entire fleet of that model, and requests Boeing to delete the obsolete configurations.

WIRING DIAGRAMS

图 5-5　线路图简介

（5）图表和清单（CHARTS AND LISTS）：如图 5-6 所示，介绍 91 章图表和清单。图表包括飞机站位布局、电线区域、主线束、面板和设备支架位置等；清单包括跳开关清单、支架清单等。

737-700
WIRING DIAGRAM MANUAL

INTRODUCTION

1. **CHARTS**

The Chapter 91 Charts contain airplane station arrangements, wire zones, major wire bundle pathways, panel and equipment shelf locations, circuit breaker panel charts, disconnect bracket charts and Master Bundle information.

2. **LISTS**

The Chapter 91 Lists are numbered as follows:

91-02-00	- -	Circuit Breaker List
91-04-00	- -	Bracket List
91-21-11	- -	Wire List
91-21-12	- -	Spare Wire List
91-21-13	- -	Master Bundle List
91-21-21	- -	Ground List
91-21-31	- -	Splice List
91-21-41	- -	Terminal Strip List
91-21-51	- -	Hookup List

The following paragraphs in this section define the contents of Chapter 91 Lists. The Wire List is the Primary source for Spare Wire through Hookup Lists.

A. **Circuit Breaker List—Chapter 91-02-00**

(a) The Circuit Breaker List reflects all the circuit breakers within an airplane and is derived from data contained in the Equipment List. It lists, in alphanumeric order, each Panel/Access Door, the Description and the Diagram of that panel.

(b) For each Panel/Access Door the grid location (Grid No), the circuit breaker number (Ckt Bkr), circuit breaker label (Description), Diagram and Effectivity are listed.

(c) Unused grid locations are not listed.

(d) The Circuit Breaker List is used as supplemental data for all Chapter 91-02-XX Panel Charts containing circuit breakers.

B. **Bracket List—Chapter 91-04-00**

(a) The Bracket List reflects all the disconnect brackets within an airplane and is derived from data contained in the Equipment List. It lists, in alphanumeric order, each disconnect bracket (BRACKET NO.), title (DESCRIPTION), maximum number of positions (MAX POS), location (STATION/WL/BL), and EFFECTIVITY.

(b) Each POSITION within a bracket is listed, followed by the mounted receptacle number (RECEPTACLE) and its wire bundle number (BUNDLE), the mating plug (PLUG) and wire bundle number (BUNDLE) and the EFFECTIVITY.

CHARTS AND LISTS

图 5-6　图表与清单简介

（6）代码（CODES）：介绍了导线类型代码、终端代码、供应商代码等，如图 5-7 所示。

737-700
WIRING DIAGRAM MANUAL

INTRODUCTION

1. **WIRE TYPE CODE**

This information is covered in Chapter 20, Standard Wiring Practices, Section 20-00-13.

2. **VENDOR CODE**

For Vendor Code translation, refer to:

H4-1: Federal Supply Code for Manufacturers-Name to Code
H4-2: Federal Supply Code for Manufacturers-Code to Name
H4-3: Nato Supply Code for Manufacturers-Name to Code/Code to Name

Published by:

Defense Supply Agency
Defense Logistics Services Center
Federal Center Building
Battle Creek, Michigan 49016

VENDOR CODE	SPECIAL VENDOR CODE ASSIGNMENT
V96906	Parts having Military Part Numbers
VAAL	American Airlines Inc. Tulsa, Oklahoma
VAB	Coastal Mfg. Co. Santa Monica, California
VAC	Safety Industries Inc. Glen Ellyn, Illinois
VAD	Glarban Corp. Gordonville, N.Y.
VAE	Ucinite Co., The Los Angeles, California
VAF	Air France 1 Square Max Hymans 75, Paris 15, France
VAI	Industrial Products Co. Gardena, California
VAJ	Bozak Sales Co. Salisbury, Connecticut
VAO	Teddington Controls Ltd. Tydfil, South Wales
VARINC	Arinc
VARMED	Airmed Ltd. Edinburgh Way Harlow, Essex, England
VAZ	Murphy Radio Ltd. Welwyn Garden City

CODES

图 5-7 代码简介

（7）手册的使用（MANUAL USAGE）：如图 5-8 所示，用来介绍线路图手册的使用方法。

737-700
WIRING DIAGRAM MANUAL

INTRODUCTION

1. **METHODS USED TO FIND INFORMATION**

   A. **How To Locate A Diagram From A Wire Found In The Airplane**

   (a) As an example, take wire number W4199-0013-24.

   (b) Knowing the wire bundle number W4199, refer to the Wire List in Chapter 91

   (c) Using Self Indexing wire list shown, locate wire bundle W4199.

   (d) Locate wire number 0013-24 and on the same line under the "Diagram" heading locate the diagram number 35-11-11.

   (e) Refer to Chapter 35 and locate the information needed on diagram 35-11-11.

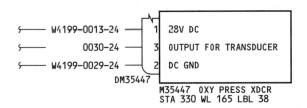

   B. **How To Locate, In The Airplane, A Wire Found On A Wiring Diagram**

   (a) Wire number W4232-0118-22 appears on Wiring Diagram 28-21-11.

   (b) Both ends of the wire are identified, disconnect D21028P at P210 panel and disconnect D31013P at P310 panel.

   (c) Another method is to locate the bundle and wire number in the Wire List. The title of the bundle usually provides Station or Area for Airframe Bundles.

MANUAL USAGE

图 5-8　手册使用方法简介

任务实施

1. 飞机有效性查询

打开前言（Front Matter），单击飞机有效性列表（Effective Aircraft）查询飞机有效性代码，如图5-9所示。

737-700
WIRING DIAGRAM MANUAL

This manual is applicable to the aircraft on this list:

| Model-Series | Operator | | Manufacturer | | | Registration Number |
	Identification Code	Effectivity Code	Block Number	Serial Number	Line Number	
737-700	XXX	001	YA811	29912	140	B-2639
737-700	XXX	002	YA812	29913	148	B-2640
737-700	XXX	003	YA813	30074	292	B-2503
737-700	XXX	004	YA814	30075	311	B-2502

图 5-9　飞机有效性列表

2. 读懂故障隔离手册的飞机有效性列表

与项目1中提到的波音飞机维护手册类似，由图5-9我们可以轻松判断，飞机 B 的有效性代码为004，造册号为 YA814。

思考题

1. 学习了本任务，你知道 WDM 的使用方法在什么地方能找到吗？
2. 相较前面我们学习过的 FIM，你能说出 WDM 的前言部分中，它们的异同点吗？

任务2　襟翼位置传感器相关线路分析

任务描述

线路图手册的有效性查询确认了飞机的有效性和适用性。接下来，我们将使用满足飞机有效性的 WDM 手册，查询自动飞行控制系统 B 中 SNSR EXC AC 跳开关所在章节。

线路图手册（WDM）与零部件图解目录手册（IPC）、飞机维护手册（AMM）、故障隔离手册（FIM）都属于客户化手册，章节划分按照《ATA 100 规范》编排。我们只要充分熟悉《ATA 100 规范》章节划分的原则，就能快速地查询并确定自动飞行控制系统所在章。

你还记得《ATA 100 规范》中，自动飞行控制系统归属哪个系统吗？借助已学过的 AMM PART I SDS（Systems Description Section），你是否能获取自动飞行控制系统 B 中

SNSR EXC AC 跳开关的相关信息？

任务要求

● 了解线路图手册（WDM）的结构。

● 了解自动飞行控制系统 B 中 SNSR EXC AC 跳开关相关信息的查询。

知识链接

1. WDM 的结构

WDM 手册的结构和 IPC、AMM、FIM 基本相同，手册包含前言和各系统章节图纸。但是，除此以外，WDM 手册还包含 00 章、91 章和设备清单。

（1）WDM 手册 00 章。WDM 手册在系统章节之前，首先是 00-00-00，该章节给出了 WDM 手册中出现的所有元器件符号（Symbol）简图。例如：如图 5 10 所示，给出了天线（ANTENNA）、电容器（CAPACITOR）、插头和插座（PLUG AND RECEPTACLE）、跳开关（CIRCUIT BREAKER）、保险丝（FUSE）等的简图。

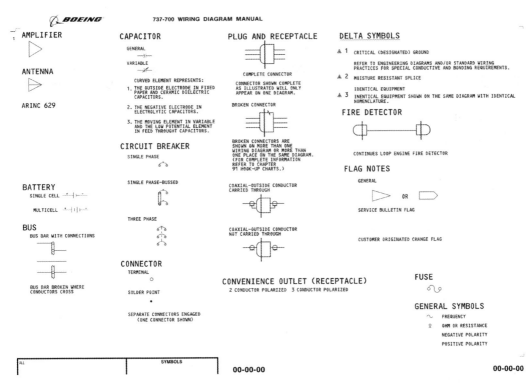

图 5-10　WDM 元器件符号简图

（2）WDM 手册 91 章。有关设备组件和设备架的位置及常用略图布局编号见表 5-1。

表 5-1　设备组件和设备架的位置及常用略图布局编号

91-00-01 ～ 91-00-99	主线束略图
91-01-01 ～ 91-01-99	常用位置略图
91-02-00 ～ 91-02-00	跳开关清单
91-02-01 ～ 91-02-99	备用
91-03-01 ～ 91-03-99	架 / 板插座接线和脱开略图
91-04-00 ～ 91-04-00	脱开支架清单
91-04-01 ～ 91-04-99	脱开支架和略图
91-05-01 ～ 91-05-99	备用
91-06-01 ～ 91-06-99	程序销钉位置图
91-07-01 ～ 91-19-99	备用

　　91 章清单详细内容见表 5-2 至表 5-3，需要重点关注的是各种清单所在的章节，例如：跳开关清单（CIRCUIT BREAKER LIST）、支架清单（BRACKET LIST）、导线清单（WIRE LIST）、备用线清单（SPARE WIRE LIST）、主线束清单（MASTER BUNDLE LIST）、接地清单（GROUND LIST）、接线管清单（SPLICE LIST）、接线端清单（TERMNIAL LIST）、连接清单（HOOK-UP LIST）等。

表 5-2　清单详细内容

章节	TABLE OF CONTENTS	具体内容
91-00		
91-00-01	WIRE BUNDLE MASTER-ELECTRONIC	主导线束 - 电子
91-00-02	WIRE BUNDLE MASTER-ELECTRICAL	主导线束 - 电气
91-01		
91-01-01	ANTENNA LOCATIONS	天线位置
91-01-02	MAJOR RACEWAYS-300	主线槽 -300
91-01-03	PANEL LOCATIONS	面板位置
91-01-04	SHELF LOCATIONS	支架位置
91-01-05	JUNCTION BOX LOCATIONS	接线盒位置
91-01-07	LOCATION OF DUAL GROUNDS INUNPRESSURIZED AREAS	非增压区的双重接地位置
91-01-08	INSTALLATION OF DUAL GROUNDS	双重接地的安装
91-02		
91-02-00	CIRCUIT BREAKER LIST	跳开关清单
91-03		
91-03-01	El-1 ELECTRONIC SHELF	El-1 电子设备架

续表

章节	TABLE OF CONTENTS	具体内容
91-03-02	E1-2 ELECTRONIC SHELF	El-2 电子设备架
91-03-03	E1-3 ELECTRONIC SHELF	El-3 电子设备架
91-03-04	E2-1 ELECTRONIC SHELF	E2-1 电子设备架
91-03-05	E2-2 ELECTRONIC SHELF	E2-2 电子设备架
91-03-06	E2-3 ELECTRONIC SHELF	E2-3 电子设备架
91-03-07	E2-4 ELECTRONIC SHELF	E2-4 电子设备架
91-03-08	E3-1 ELECTRICAL SHELF	E3-1 电子设备架
91-03-09	E3-2 ELECTRICAL SHELF	E3-2 电子设备架
91-03-10	E3-3 ELECTRICAL SHELF	E3-3 电子设备架
91-03-11	E3-4 ELECTRONIC SHELF	E3-4 电子设备架
91-03-12	Pl CAPTAINS INSTRUMENT PANEL	P1 正驾驶仪表板
91-03-13	P2 CENTER INSTRUMENT PANEL	P2 中央仪表板
91-03-14	P3 FIRST OFFICERS INSTRUMENT PANEL	P2 副驾驶仪表板
91-03-15	P5 FORWARD OVERHEAD PANEL	P5 前顶板
91-03-16	P5 AFT OVERHEAD PANEL AND DISCONNECT PANELS	P5 后顶板和脱开面板
91-03-17	P6 MAIN BREAKER PANEL LAYOUT AND TERMINAL STRIPS	P6 主跳开关面板布局和接线条
91-03-18	P6-l PANEL	P6-1 板
91-03-19	P6-2 PANEL	P6-2 板
91-03-20	P6-3 PANEL	P6-3 板
91-03-21	P6-4 PANEL	P6-4 板
91-03-22	P6-5 AND P6-6 PANELS	P6-5 板和 P6-6 板
91-03-23	P6-11 AND P6-12 PANELS	P6-11 板和 P6-12 板
91-03-24	P6 EQUIPMENT	P6 设备
91-03-25	P18 LOAD CONTROL CENTER-LEFT, PANEL LAYOUT AND TERMINAL STRIP	P18 左侧负载控制中心面板布局和接线条
91-03-26	P18-1 AND P18-2 PANELS	P18-1 板和 P18-2 板
91-03-27	P18-3，P18-4AND P18-5 PANELS	P18-3 板、P18-4 板 和 P18-5 板
91-03-29	P8 AND P9 FORWARD AND AFT ELECTRONIC PANELS	P8 和 P9 前、后电子控制面板
91-03-34	E6-1 ELECTRONIC SHELF	E6-1 电子设备架
91-03-37	E3-5 ELECTRONIC SHELF	E3-5 电子设备架
91-04	DISCONNECT BRACKET LIST	脱开支架清单

章节	TABLE OF CONTENTS	具体内容
91-04-00	BRACKET LIST	支架清单
91-04-01	DISCONNECT BRACKETS LOCATIONS	脱开支架位置
91-04-02	MAIN INTRUMENT PNL，AISLE CONTROL STAND AND E11 DISCONNECT BRACKETS	主仪表板、走廊控制板和E11 脱开支架
91-04-03	P5, P6 AND P18 PANEL DISCONNECT BRACKET	P5、P6 和 P18 板脱开支架
91-04-04	E1 SHELF DISCONNECT BRACKETS	E1 设备架脱开支架
91-04-05	E2 SHELF DISCONNECT BRACKETS	E2 设备架脱开支架
91-04-06	E3 SHELF DISCONNECT BRACKETS	E3 设备架脱开支架
91-04-07	PANEL DISCONNECT BRACKETS	面板脱开支架
91-04-08	MAIN WHEEL WELL，L&R WING AND ENGINE FIREWALL DISCONNECT BRACKETS	主轮舱、左右大翼和发动机防火墙脱开支架
91-04-09	AFT BODY AND E6 SHELF DISCONNECT BRACKETS	后机体和 E6 架脱开支架
91-04-10	AFT BODY AND E5 SHELF DISCONNECT BRACKETS	后机体和 E5 架脱开支架
91-06	PROGRAM PIN	程序销钉
91-06-01	PROGRAM PIN SWITCH/ CODING SWITCH LOCATIONS	程序销钉译码电门位置

表 5-3　清单表

91-02-00	CIRCUIT BREAKER LIST	跳开关清单
91-04-00	BRACKET LIST	支架清单
91-21-11	WIRE LIST	导线清单
91-21-12	SPARE WIRE LIST	备用线清单
91-21-13	MASTER BUNDLE LIST	主线束清单
91-21-21	GROUND LIST	接地清单
91-21-31	SPLICE LIST	接线管清单
91-21-41	TERMNIAL LIST	接线端清单
91-21-51	HOOK-UP LIST	连接清单

（3）WDM 手册系统线路图。WDM 手册 21 章至 80 章给出了系统章节线路图纸。WDM 系统章节图纸示例如图 5-11 所示。

2. 襟翼位置传感器的相关知识

飞机自动飞行控制系统 B 中，造成 SNSR EXC AC 跳开关无法复位故障的原因很多。在故障维护过程中，必须先明确具体的故障现象。根据任务描述，本项目故障现象为自动飞行控制系统中襟翼位置传感器故障（Flap Position Transmitter Problem）。为了更好地理解本项目的故障现象，下面简单介绍飞行控

襟翼位置传感器的相关知识

制系统以及飞机襟翼系统的相关知识。WDM 系统线路图纸示例如图 5-11 所示。

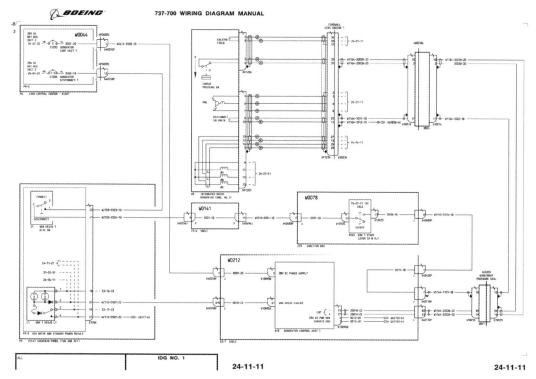

图 5-11　WDM 系统线路图纸示例

　　我们都知道，飞行员执行飞行任务时，通过飞行控制系统完成相关操作，系统功能正常与否不仅影响到飞机的气动外形、飞行品质、乘坐体验，更直接关系到飞行安全。飞机的横滚、俯仰、偏航等功能均需通过飞行操纵系统得以实现。而襟翼系统（Flap Control System）作为飞行控制系统的重要组成部分，分为前缘装置（LE FLAPS、LE SLATS）和后缘装置（TE FLAPS）。后缘襟翼位置传感器位于左右后缘襟翼的扭力管处，其作用是精确记录后缘襟翼的位置信息并传至襟翼 / 缝翼电子组件（Flap/Slat Electronics Unit，FSEU），FSEU 比较两侧数值，如果差值超过 9°，会触发襟翼不对称保护，严重的情况下会引起复飞。因此，排除后缘襟翼位置传感器故障对保障飞行安全具有重要。

任务实施

1. 飞机 SNSR EXC AC 跳开关相关信息的查询

飞机 SNSR EXC AC
跳开关相关信息的查询

　　根据任务描述，飞机自动飞行控制系统 B 中，SNSR EXC AC 跳开关无法复位，且故障现象为襟翼位置传感器故障（Flap Position Transmitter Problem）。查询故障隔离手册（FIM）22 章自动飞行控制系统（AUTO FLIGHT）可获得该故障可能的原因及故障隔离程序。

如图 5-12 所示，故障可能的原因（Possible Causes）依次为：左缘襟翼传感器 T427 故障；右缘襟翼传感器 T428 故障；飞行控制计算机 A（FCC-A）M1875 故障；飞行控制计算机 B（FCC-B）M1876 故障；线路故障（Wiring Problem）。

BOEING®
737-600/700/800/900
FAULT ISOLATION MANUAL

807. Flap Position Transmitter Problem - Fault Isolation

  A. Description

    (1) This task is for these maintenance messages:

    <u>NOTE</u>: J1A-XX, J1B-XX and J1C-XX (where XX is a pin number) are the connector and pin numbers on the FCC-A and FCC-B. For the FCC-A: J1A = D10135A, J1B = D10135B and J1C = D10135C. For the FCC-B: J1A = 10137A, J1B = 10137B and J1C = 10137C.

      (a) FLAP POSN-1 (J1B-G06, F06, E06)

      (b) FLAP POSN-2 (J1B-G06, F06, E06)

    (2) FLAP POSN-1 (J1B-G06, F06, E06): The flight control computer A (FCC-A) receives incorrect data from the left flap position transmitter during the ground test.

    (3) FLAP POSN-2 (J1B-G06, F06, E06): The flight control computer B (FCC-B) receives incorrect data from the right flap position transmitter during the ground test.

  B. Possible Causes

    (1) Left flap position transmitter, T427

    (2) Right flap position transmitter, T428

    (3) Flight control computer A (FCC-A), M1875

    (4) Flight control computer B (FCC-B), M1876

    (5) Wiring Problem

  C. Circuit Breakers

    (1) These are the primary circuit breakers related to the fault:

    CAPT Electrical System Panel, P18-1

Row	Col	Number	Name
C	5	C01041	AFCS SYS A SNSR EXC AC

    F/O Electrical System Panel, P6-2

Row	Col	Number	Name
C	2	C01042	AFCS SYS B SNSR EXC AC

  D. Related Data

    (1) (SSM 22-11-11)

    (2) (SSM 22-12-31)

    (3) (SSM 22-12-41)

    (4) (WDM 22-11-11)

    (5) (WDM 22-12-31)

    (6) (WDM 22-12-41)

EFFECTIVITY
XXX ALL

22-12 TASK 807

图 5-12　故障相关信息

　　故障相关跳开关（Circuit Breakers）包括自动飞行控制系统 A（Automatic Flight Control System，AFCS）SNSR EXC AC 跳开关 C01041 和自动飞行控制系统 B SNSR EXC AC 跳开关 C01042。

　　相关手册资料（Related Data）有系统原理图手册（SSM）和线路图手册（WDM），章节锁定在 22 章自动飞行控制系统。

　　如图 5-13 所示，故障隔离程序给出飞机自动飞行控制系统 B 中，SNSR EXC AC 跳开关无法复位故障，相关的跳开关为 C01042，需检测右缘襟翼位置传感器 T428 的连接器 D229 的导通性。

(4) If the maintenance message is on channel B, then do these steps:

    (a) Disconnect the right flap position transmitter connector, D229.

        NOTE: You can find the left flap position transmitter on the outboard end of the transmission No. 8.

    (b) Do a continuity check between the pins shown below, at the flap position transmitter, T428.

D229		D229
pin 9	------------------	pin 10
pin 12	------------------	pin 13
pin 12	------------------	pin 14
pin 13	------------------	pin 14

    (c) If there is an open circuit, do these steps;

        1) Replace the flap position transmitter, T428.

        These are the tasks:

        Flap Position Transmitter Removal, AMM TASK 27-58-01-000-801,

        Flap Position Transmitter Installation, AMM TASK 27-58-01-400-801.

        2) Do the Repair Confirmation at the end of this task.

    (d) If there is continuity between the above pins, then continue.

    (e) Make sure that this circuit breaker is closed:

    F/O Electrical System Panel, P6-2

Row	Col	Number	Name
C	2	C01042	AFCS SYS B SNSR EXC AC

图 5-13　故障隔离程序

2. 飞机右缘襟翼位置传感器相关线路分析

飞机右缘襟翼位置传感器
相关线路分析

　　打开 WDM 手册 22 章自动飞行控制系统（AUTO FLIGHT）的 22-11-11，查询 D229 的 Pin9、Pin10、Pin12、Pin13、Pin14 的相关线路，如图 5-14 所示。

　　由 WDM 22-12-41 可知，D229 的 Pin9 通过导线 W1034-2005R-20 与 D40034J 的 Pin24 相连，D229 的 Pin10 通过导线 W1034-2005B-20 与 D40034J 的 Pin25 相连。

思考题

1. 你能查询并记录 D229 的 Pin12、Pin13、Pin14 的相关线路信息吗？

2. 请你说说 WDM 22-12-41 中电源及跳开关的相关信息。

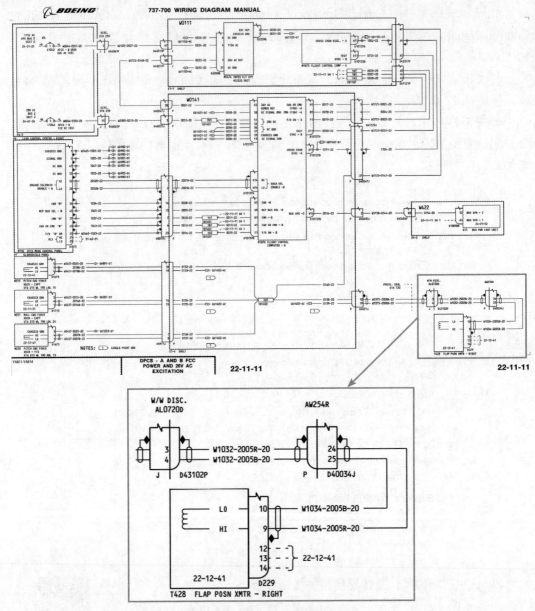

图 5-14 襟翼位置传感器线路图

任务 3　襟翼位置传感器线路故障的排除

任务描述

已知飞机自动飞行控制系统 B 中襟翼位置传感器（Flap Position Transmitter）连接

器插钉（Pin）线路信息，需按照故障隔离程序排除线路故障。上一任务中，我们已经知晓了襟翼位置传感器相关线路所在章节，为了制定其线路故障排除方案，我们还需要以 WDM 22-12-41 线路图为契机，借助 WDM 强大的清单，深入探寻右缘襟翼位置传感器 T428 的连接器 D229 的连线信息，排除襟翼位置传感器的线路故障。

如何使用 WDM 手册查询设备清单、导线清单、跳开关清单等？查询清单时可以额外获得哪些重要信息？

任务要求

- 掌握 WDM 手册中各式清单的使用。
- 掌握故障排除工卡的编制。

知识链接

1. WDM 手册常用清单

（1）设备清单（Equipment List，EQ）。如图 5-15 所示，WDM 手册的设备清单给出了设备号（Equipment Number）、选装说明（Option）、件号（Part Number）、件的描述（Part Description）、安装细节图（Used on Drawing）、供应商代码（Vendor Code）、安装数量（Quantity）、线路图章节号（Diagram Field）、定位（Location）、有效性（Effectivity）。这些信息在 WDM 手册前言的设备清单中均有详细介绍，如图 5-4 所示。

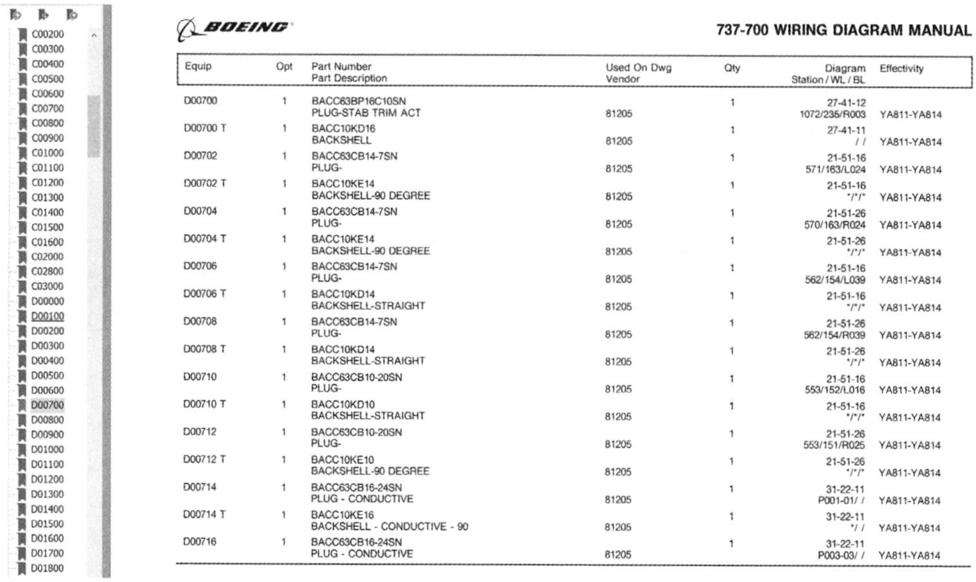

BOEING　　737-700 WIRING DIAGRAM MANUAL

Equip	Opt	Part Number Part Description	Used On Dwg Vendor	Qty	Diagram Station / WL / BL	Effectivity
D00700	1	BACC63BP16C10SN PLUG-STAB TRIM ACT	81205	1	27-41-12 1072/235/R003	YA811-YA814
D00700 T	1	BACC10KD16 BACKSHELL	81205	1	27-41-11 / /	YA811-YA814
D00702	1	BACC63CB14-7SN PLUG-	81205	1	21-51-16 571/163/L024	YA811-YA814
D00702 T	1	BACC10KE14 BACKSHELL-90 DEGREE	81205	1	21-51-16 */*/*	YA811-YA814
D00704	1	BACC63CB14-7SN PLUG-	81205	1	21-51-26 570/163/R024	YA811-YA814
D00704 T	1	BACC10KE14 BACKSHELL-90 DEGREE	81205	1	21-51-26 */*/*	YA811-YA814
D00706	1	BACC63CB14-7SN PLUG-	81205	1	21-51-16 562/154/L039	YA811-YA814
D00706 T	1	BACC10KD14 BACKSHELL-STRAIGHT	81205	1	21-51-16 */*/*	YA811-YA814
D00708	1	BACC63CB14-7SN PLUG-	81205	1	21-51-26 562/154/R039	YA811-YA814
D00708 T	1	BACC10KD14 BACKSHELL-STRAIGHT	81205	1	21-51-26 */*/*	YA811-YA814
D00710	1	BACC63CB10-20SN PLUG-	81205	1	21-51-16 553/152/L016	YA811-YA814
D00710 T	1	BACC10KD10 BACKSHELL-STRAIGHT	81205	1	21-51-16 */*/*	YA811-YA814
D00712	1	BACC63CB10-20SN PLUG-	81205	1	21-51-26 553/151/R025	YA811-YA814
D00712 T	1	BACC10KE10 BACKSHELL-90 DEGREE	81205	1	21-51-26 */*/*	YA811-YA814
D00714	1	BACC63CB16-24SN PLUG - CONDUCTIVE	81205	1	31-22-11 P001-01/ /	YA811-YA814
D00714 T	1	BACC10KE16 BACKSHELL - CONDUCTIVE - 90	81205	1	31-22-11 */ /	YA811-YA814
D00716	1	BACC63CB16-24SN PLUG - CONDUCTIVE	81205	1	31-22-11 P003-03/ /	YA811-YA814

图 5-15　设备清单

Part Number：设备的件号。

Part Description：设备的名称。

Used On Dwg（Used on Drawing）：该部件所在的图纸，图纸包含了该部件的详细安装情况，主要供波音公司使用，目的在于通过该图号把所有与此设备相关的图纸联系起来，最后便于得出飞机的整图。

Vendor（Vendor Code）：生产此设备的供应商代码表示。

Qty（Quantity）：表示该部件在飞机上安装的个数。

Diagram（Diagram Field）：该部件所在的线路图章节号；有的部件会显示多个线路图。

Station/WL/BL（Location）：部件的定位信息，Station/WL/BL 为站位 / 水线 / 纵剖线。若部件位于或临近面板、设备架、脱开支架、终端块，则直接用它们的设备号表示。

Effectivity：有效性。通过有效性代码确定设备号的适用性。代码可以是波音的有效性代码，也可以是用户自己规定的代号，由用户自己选择。一个部件有效性代码可以是一架飞机或某个范围，如 PP001 ～ PP099，其中包括了很多架飞机，ALL 表示对该手册所属的所有飞机有效。

其中，有两个内容需要额外强调和说明。

首先，关于设备号（Equipment Number），飞机上所有的设备号都由首位字母加数字组成，编排规则为一位字母加五位数字，或者两位字母加五位数字，不够在前面补零。例如：D700 在设备清单中写为 D00700，切不可误写为 D70000。以此类推，AB372 在设备清单中写为 AB0372。插头 D00700 件号为 BACC63CBP16C10SN，其后壳（BACKSHELL）D00700T 的件号为 BACC10KD16。

其次，关于选装说明（Option），此区域表示是否可使用其他件号的设备。如表 5-4 所列，Boeing "as delivered" Options 为波音公司交付选项，当波音选装件此区域为空白或为 0 时，表示不允有选择；此区域为 1 时，表明选项可用。Customer Requested Options 为飞机交付后，根据客户需求选装，当有多个选项可用时，允许四个选项。

表 5-4　选装说明

Boeing "as delivered" Options		Customer Requested Options	
OPT	Option available	OPT	Option available
	No	9	First option
0	No	8	Second option
1	First option	7	Third option
2	Second option	6	Fourth option
3	Third option		

（2）跳开关清单（Circuit Breaker List）。跳开关清单按跳开关所在面板或接线盒的字母顺序排列，包含了飞机上所有的跳开关。如图 5-16 所示，跳开关清单给出了它的设备号、描述、线路图章节号、有效性，以及所在的面板号（Panel）、位置号（Grid Number）等信息。

737-700 WIRING DIAGRAM MANUAL

Panel/Access Door		Description		Diagram		
	Grid No	Ckt Bkr	Description		Diagram	Effectivity
J009-00						ALL
	A004	C01341	CB-STATIC INVERTER RCCB		24-34-11	ALL
	A005	C00142	CB-BATTERY CHARGER		24-31-11	ALL
		C01340	CB-BATTERY BUS		24-31-12	ALL

图 5-16　跳开关清单

Panel No：跳开关所在的面板号。

Grid No：跳开关所在面板上的位置号。

Ckt Bkr：跳开关的设备号。

Description：跳开关的具体描述，说明此跳开关的用途。

Diagram：跳开关所在的线路图章节号。

Effectivity：有效性。

（3）脱开支架清单（Disconnect Bracket List）。如图 5-17 所示，通过脱开支架清单可以查阅到脱开支架的设备号、所在位置、定位信息和可供放置插头 / 座的最大位置数量、每一个位置上的插头 / 座设备号，以及该插头 / 座连接的线束号等信息。

737-700 WIRING DIAGRAM MANUAL

Bracket No.	Description					Max Pos	Station / WL / BL	Effectivity
	Position	Receptacle	Bundle	Plug	Bundle			
AB0237A	BRACKET					000	237/201/L004	ALL
	01	D11160	W5162					YA812-YA830
	02	D11158	W5172					YA812-YA830
	001 UNUSED							YA811
	002 UNUSED							YA811

图 5-17　脱开支架清单

Bracket No：脱开支架的设备号。

Position：脱开支架上每一个插头 / 座的位置号。

Description：脱开支架的具体描述。

Receptacle：插座设备号，脱开支架上对应 Position 位置的插座设备号。

Bundle：插座连接的线束号。

Plug：插头设备号，脱开支架上对应 Position 位置与插座相连的插头设备号。

Bundle：插头连接的线束号。

Max Pos：脱开支架上可供放置插头 / 座的最大位置数量。

（4）导线清单（Wire List）。导线清单以导线束号顺序排列，例如 W****，导线清单中包括导线束号（Bundle No）、导线束件号（Part Number）、线径（Gauge）、类型（Type）、导线的描述、导线所连接的设备等信息。

Bundle No：导线束号，导线所在的导线束号。

Part Number and Description：导线束件号和描述，对此线束的功能进行简单说明。

WIRE No./GA/CO Field（Wire Number/Gauge/Color）：导线的序列号和规格，表明

导线的类型、颜色等（颜色代码也可能出现在线号中）。每根导线上都打印或压印有线号，也可以根据线路图来识别导线。

导线编号的描述一般由三部分组成，即导线束号、线号、线径，格式为 W****-****-**。如图 5-18 所示，导线编号为 W2201-0021-20 的导线类型代码为 PA。

BOEING 　　　　　　　　　　　　　　　　　　　**737-700 WIRING DIAGRAM MANUAL**

Bundle No. Wire No.	Part Number	GA	CO	TY	Fam	Description FT-IN	Diagram	From Equip	Term	Type	Splice	To Equip	Term	Type	Splice	Effectivity
W2201	286A2201					LEFT AC ELEX PWR FROM P18 TO E3. (continued)										
0021		20	PA			23-0	SPARE	D41805P	21			D41817P	5			YA811-YA812
0021		20	PA			23-6	SPARE	D41805P	21			D41817P	5			YA813-YA814
0025		22	PA			24-10	23-27-13	D41805P	25			D43313P	5			YA811-YA812
0025		22	PA			25-5	23-27-13	D41805P	25			D43313P	5			YA813-YA814
0026		22	CQ			24-10	23-27-11	D41805P	26			D43313P	6			YA811-YA812
0026		22	CQ			25-5	23-27-11	D41805P	26			D43313P	6			YA813-YA814

图 5-18　导线清单

其中，表 5-5 给出导线序列号，901～999 和 9001～9999 预留给客户使用。

表 5-5　导线序列号表

导线种类	导线序列号
单股线（不带屏蔽层）	001～199
单股线（带屏蔽层） 屏蔽层	001～199 SHA～SPZ
双股线（不带屏蔽层）	201～299
双股线（带屏蔽层） 屏蔽层	201～299 SHA～SPZ
三股线（不带屏蔽层）	301～399
三股线（带屏蔽层） 屏蔽层	301～399 SHA～SPZ
四股线（不带屏蔽层）	401～450
四股线（带屏蔽层） 屏蔽层	401～450 SHA～SPZ
特殊导线（耐高温、热电偶用等）	451～500
其他导线（不包括屏蔽层）	501～899 8000～8999
留用线号	900～999 9000～9999
同轴电缆（带屏蔽层） 屏蔽层	451～500 SHA～SPZ
双股线（带屏蔽层） 屏蔽层	451～500 SHA～SPZ

续表

导线种类	导线序列号
三股电缆 内层屏蔽 外层屏蔽	451 ～ 500 SHI SHO
多余 900 根导线的线束（A 和 W 可以与数字以任意方式组合，但不能使用其他字母）	01A ～ A99 01W ～ W99
厂家配置的导线 飞机上保留线号	A-A ～ Z-Z 9-1 ～ 9-99
套管 工程用 模型用	 S01 ～ S99 P01 ～ P99
编织型线束	YAA ～ YZZ
继电器定位销 飞机上保留线号	PA1 ～ PY9 PZI ～ PZZ
总线汇流条 接线片用 线路跳开关绝缘用	 ZZA ～ ZZ9 YAA ～ YZZ

CO（Color）：导线的颜色代码。导线分为单股导线和多股导线，单股导线颜色代码表示方法见表 5-6，多股导线颜色代码表示方法见表 5-7。

表 5-6 单股导线的颜色代码

导线颜色	代码	导线颜色	代码	导线颜色	代码	导线颜色	代码	导线颜色	代码
灰	A	蓝	B	绿	G	黑	K	白	W
红	R	褐	T	紫	Y				
不带绝缘层	U								

表 5-7 多股导线的颜色代码

导线颜色	代码	导线颜色	代码	导线颜色	代码	导线颜色	代码	导线颜色	代码
黑 / 蓝	KB	蓝 / 黑	BK	红 / 黑	RK	白 / 黑	WK	白 / 黑 / 红	AG
黑 / 棕	KN	蓝 / 棕	BN	红 / 蓝	RB	白 / 蓝	WB	白 / 黑 / 紫	AL
黑 / 绿	KG	蓝 / 绿	BG	红 / 棕	RN	白 / 棕	WN	白 / 黑 / 黄	AM
黑 / 灰	KA	蓝 / 黄	BY	红 / 绿	RG	白 / 灰	WA	白 / 蓝 / 黑	AK
黑 / 橙	KO	蓝 / 橙	BO	红 / 橙	RO	白 / 绿	WG	白 / 蓝 / 黄	AA
黑 / 红	KR	蓝 / 粉	BP	红 / 粉	RP	白 / 红	WR	白 / 棕 / 红	AH
黑 / 紫	KV	蓝 / 紫红	BL	红 / 紫红	RL	白 / 紫	WV	白 / 绿 / 红	AJ
黑 / 黄	KY	蓝 / 绿 / 黑	AC	红 / 紫	RV	白 / 黄	WY	白 / 红 / 蓝	AD

导线颜色	代码	导线颜色	代码	导线颜色	代码	导线颜色	代码	导线颜色	代码
				红 / 黄	RY	白 / 橙	WO	白 / 黄 / 红	AE
橙 / 黑	OK	黄 / 黑	YK	红 / 黄 / 绿	AB	白 / 粉	WP	白 / 紫红 / 红	AF
橙 / 棕	ON	黄 / 绿	YG			白 / 紫红	WL		
橙 / 黄	OY	黄 / 橙	YO	棕 / 橙	NO				
		黄 / 紫	YL						

TY（Wire Type）：导线类型代码。通过此代码与 20-00-13 导线类型码（Wire Type Codes）相互联系起来，用来查找导线件号。具体内容详见 SWPM 相关项目。

Fam（Wire Family）：组代码。如表 5-8 所列，组代码用来表示相互关联的一组线，比如有共同的屏蔽层或绝缘层或者是扭曲在一起的导线。

表 5-8　组代码表

导线种类	组代码
导线	A ～ Z、AA ～ LZ
同轴电缆、同轴轴芯、同轴双层屏蔽电缆	MA ～ MZ
复杂的导线组、组内细分小组、其第一个数字应相同	1A ～ 9Z

FT-IN（Wire Length）：长度，英尺 / 英寸，表示导线的长度。在线路图中会标识出导线或线束的关键长度和允许误差。若为空白，表示它是组线中的一部分，其长度在组线的最小序列号的导线中显示。

Diagram（Diagram Reference）：导线所在的线路图章节号。MU-LT-I 表示此导线同时出现在几张线路图上。SP-AR-E 表示此导线是一根备用线。如客户启用备用线，应在手册中修改该导线。

Equip From and To（Equipment Number）：两个设备号，表示与导线两端相连接的设备。除了接地点（GD****）和连接管（SP****、SM****），这些设备均可在设备清单中查询。若两设备号相同，则表示导线是从一个设备的一点连接到同一设备的另一点。如图 5-18 所示，编号为 W2201-0021-20 的导线连接 D41805P 的 21 号钉与 D41817P 的 5 号钉（注意有效性）。

Term From and To（Terminal Number）：两个终端号。表示与导线两端相连的终端，用字母数字表示。

标识前加 "=" 号，仅用于连接及测试目的，在部件上找不到该标识。REF 用于表示一个邻近的终端，但不表示在该设备的接线管和接线端范围内；DED 表示屏蔽线的一端衔空，没有用跳线或猪尾巴线连接；CAP 表示该线的一端使用终端帽。

接地终端标识：A 为交流接地；D 为直流接地；S 为防静电、屏蔽或特殊接地。

大写字母前加 "-" 表示与大写字母相对应的小写字母。例如：-F=f。

Type Field From and To（Terminal Type）：终端类型，用代码表示。终端类型在 WDM 手册前言（Front Matter）的 CODES 中可查询。如图 5-19 所示，接线端和接地桩的尺寸和 / 或件号（在线路图中用符号来表示终端尺寸）、特殊终端或特殊插钉。

终端代码的表现形式如下：

● 单个字母。
● 以字母开头的两位字母数字。
● 数字（一位数字、两位数字、一位数字 + 一位字母）。
● 符号。

3. TERMINAL INFORMATION

The following index of "Term Type Codes" lists the code, a description of the code and, as applicable, the terminal stud size and/or part number. The code index is arranged in the following order:

● Single alphabetical letter
● Two character code with leading alphabetical letters
● Numbers
● Symbols

See Standard Wiring Practices (Chapter 20) for maintenance or repair information.

A. Single alphabetical letter

TERM TYPE CODE	DESCRIPTION OF THE CODE	PART NUMBER
A	General Purpose Lug, Standard/Narrow,	BACT12AC43

图 5-19　终端类型代码

Splice Field From and To（Splice）：用来表示拼接管的连接方式。需要说明的是，如图 5-20 所示，"*" 表示两根或多根线的终端连接在一起；而 "*1"、"*2"、"*3" 表示 "*1" 的线连接在一个终端上，"* 2" 的线连接在第二个终端上，"*3" 的线连接在第三个终端上；A，B，…表示屏蔽线终端；JPA，JPB，…表示从拼接管到另一终端的导线；FA 表示屏蔽接地是用金属箍或猪尾巴套固定的；FR-（）跳线号用于将屏蔽线连到插钉或地。例如：FRAA 表示跳线连到 FA；FRAB 表示跳线连到 FB 等。

图 5-20　拼接管的连接方式

（5）备用线清单（Spare Wire List）。备用线清单包括飞机上所有的备用线。如图 5-21 所示，备用线清单提供备用线起始的设备、设备上的位置、线束号、导线号以及备用线终端设备的设备号、位置及类型。

（6）接地清单（Ground List）。接地清单反映了两种类型的接地，GB 仅用于增压区，GD 可用于增压区或非增压区。由于支架或面板上的接地（GDM、GDX、GDY、GDZ、

GBX、GBY、GBZ）在一个特定的支架或面板上是唯一的，因此不在此清单中。接地清单中包括接地块的设备号以及接地类型（AC、DC、S）。AC 为交流接地；DC 为直流接地；S- 为防静电、屏蔽或特殊接地。

737-700 WIRING DIAGRAM MANUAL

From Matewith From	Pos	Bundle	Sep	Description		To					
Equip	Term	Type	Wire No.	GA	TY	Equip	Term	Type	Matewith	Pos	Effectivity
AB1050L	**02**	**W3393**	**LG2**	**DISCONNECT BRACKET-SEC 48 L**							
D43393J	15		1015	20	VF	D00545	15		M00414		YA811-YA814
D43393J	S/R		1015Z			GD04809	ST..	E			YA811-YA814
D43393J	17		1017	20	VF	D00545	17		M00414		YA811-YA814
D43393J	S/R		1017Z			GD04809	ST..	E			YA811-YA814
D43393J	19		1019	20	VF	D00545	19		M00414		YA811-YA814
D43393J	S/R		1019Z			GD04809	ST..	E			YA811-YA814
D43393J	21		1021	20	VF	D00545	21		M00414		YA811-YA814
D43393J	S/R		1021Z			D00545	S/R		M00414		YA811-YA814

图 5-21　备用线清单

此外，接地清单中还包括接地块定位信息、终端类型、线束号、导线的线号规格颜色、线路图章节参考号和有效性，如图 5-22 所示。

737-700 WIRING DIAGRAM MANUAL

Ground No.		Part Number				Station / WL / BL		
Term	Type	Bundle	Wire No.	GA	CO	Diagram	Effectivity
GD01748	DC					/ /	ALL
GD01753	DC-					196/171/R012	ALL
DC..	E	W5156	5510	20		21-27-21	YA811-YA814
DC..	E	W5503	0001	20		34-21-22	ALL

图 5-22　接地清单

（7）拼接管清单（Splice List）。拼接管清单反映了飞机内的全部拼接管（Splice）的信息。SP**** 用于表示连接同一导线束的拼接管。SM**** 表示用于表示连接不同导线束导线的拼接管，不在此清单内，因为该清单用于同一线束内导线的连接。

拼接管清单中包括接线管的设备号、有效性、定位信息、接线管两端的线号、终端类型及线路图章节参考号，如图 5-23 所示。

737-700 WIRING DIAGRAM MANUAL

Splice No.	Bundle	Station / WL / BL Wire No.	GA	CO	Type	Diagram	Effectivity
SP00300		**237/208/L036**					**ALL**
	W2030	1001	18		S	33-11-31	ALL
	W2030	1001Z			S	33-11-31	ALL
	W2363	1062	20		S	33-11-31	YA811-YA814
	W2363	1062Z				33-11-31	YA811-YA814

图 5-23　拼接管清单

（8）连接清单（Hook-Up List）。连接清单反映设备所有插钉的连线情况，但不包括接地点、拼接管及单相跳开关。连接清单列举了和部件连接的设备、终端类型、导线束、导线号、线径和颜色以及所在的线路图章节参考号。如图 5-24 所示，设备号为 D00512 的插头（PLUG）有 5 个插钉，其中 2、4、5 号钉没有使用（UNUSED），1 号钉连接的导线编号为 W6214-5531-20，其线路图号为 21-61-21，有效性适用于造册号 YA811 到 YA814。

737-700 WIRING DIAGRAM MANUAL

Equip	Term	Station / WL / BL Type	Bundle	Description Wire No.	GA	CO	Diagram	Effectivity
D00512		372/195/L055		PLUG				YA811-YA814
	1		W6214	5531	20		21-61-21	YA811-YA814
	2	UNUSED						ALL
	3		W6214	5530	20		21-61-21	YA811-YA814
	4	UNUSED						ALL
	5	UNUSED						ALL

图 5-24 连接清单

2. 飞机线路故障排除的相关知识

细心保航——东航"90"后
女机务年复百遍"规定动作"

飞机线路故障具有较强的隐蔽性，其分布范围广泛，因此对于故障处理和维护提出了更高的要求。分析飞机线路故障类型，有针对性地采用故障处理措施，有助于提高故障维修效率，确保飞机的安全、稳定运行。飞机线路容易出现多种类型的故障：第一，接触不良，飞机线路电线插头出现松动和老化，使得线路接触不良，影响飞机的安全运行；第二，线路断路，飞机线路往往受到接线头焊接不牢、外力等因素的影响，使得飞机线路断开；第三，线路短路，飞机线路绝缘层受到湿度、温度、外力等因素的影响而出现损坏，导致飞机机身和暴露线路相接触，出现短路故障。

（1）飞机线路故障类型和原因。飞机线路出现运行故障主要由于以下几个原因：

1）线路质量不合格。生产厂家在加工生产电线时，由于质量意识淡薄或者个人疏忽，造成电线出现缺陷或者问题，无法满足飞机应用要求。飞机线路常见的故障主要包括线路屏蔽层缺陷、冷接头缺陷等，这些缺陷会增加飞机线路电阻，影响飞机的正常运行。

2）线路设置不合理。飞机线路规划设计不科学，线路铺设过程中一些工作人员人为疏忽，使得飞机线路布置间距不符合要求，线路设置不合理。

3）线路老化。随着飞机的飞行时间和距离的不断增长，飞机线路绝缘层容易受到多种因素的影响，使得金属丝暴露，飞机无法正常地接收和发射信号。

4）连接设备腐蚀和氧化。飞机运行环境恶劣，飞机上的连接设备往往受到高压的影响，发生腐蚀或者被氧化，使得飞机线路电阻不断增大，在连接设备上产生较大压降，影响飞机线路的稳定性，干扰信号正常发射。

5）运行维护不达标。当飞机线路发生故障时，一些维护检修人员没有严格按照飞机修理手册操作和维修不规范，使得飞机线路运行维护不达标，而反复的维护检修将严重影响飞机线路的稳定性。

（2）飞机线路故障检测方法。飞机线路故障检测要充分考虑电子系统性能、线路布置、飞机机型等因素，对于不同的飞机线路故障，可采用不同的检测方法。

1）目视检测。目视检测也称为外观检测，主要是运行维护人员利用肉眼来观察飞机线路情况，特别是飞机线路布置、线路插头等情况，仔细检查插头是否松动，绝缘层是否完整或者有裂痕。这种检测方法的准确性较低，无法快速、准确地发现飞机线路故障。

2）量具检测。量具检测，顾名思义，运用简单的测量工具，如万用表、欧姆表、兆欧表等，仔细检测飞机线路的运行状态。重点检测飞机线路的阻值变化、线路是否完整或者断开等。维护检修人员在检测和维修飞机线路时主要依据维修手册，如故障隔离手册、线路图手册等，结合检测数据，确定飞机线路的故障区域和原因，制定线路故障排除方案。同时，使用万用表时，对于飞机线路的断开故障，若存在虚接情况，必须再使用欧姆表准确判断飞机线路是否断开；若飞机线路绝缘层被损坏，应注意检测大地与飞机线路之间的绝缘电阻。工作人员应严格按照飞机线路检测安全原则，规范自己的操作。由于飞机线路检测电阻较高，如检测过程中需使用兆欧表，应及时切断飞机线路电源，放掉检测设备的感应负荷，避免维修检测人员发生触电安全事故。

任务实施

襟翼位置传感器相关
导线类型代码的查询

1. 襟翼位置传感器相关飞机导线类型代码查询

本项目任务 2 已确定 WDM 手册 22-12-41 可查 Pin9 通过导线 W1034-2005R-20 与 D40034J 的 Pin24 相连，D229 的 Pin10 通过导线 W1034-2005B-20 与 D40034J 的 Pin25 相连。

打开导线清单（Wire List），如图 5-25 所示，查询导线束 W1034 中线号为 2005B、线径为 20 的导线，除了可以得到连接设备的信息，还可以获得导线类型代码（Type）为 Y6。

BOEING — 737-700 WIRING DIAGRAM MANUAL

Bundle No. Wire No.	GA	CO	TY	Fam	FT-IN	Diagram	From Equip	Term	Type	Splice	To Equip	Term	Type	Splice	Effectivity
W1032		286A1032			R WING TRAILNG EDGE:DISC BRACKET STA 254R TO W/W DISC & AFT CARGO DISC (continued)										
3011R	20		Y7	WK	0-0	27-62-14	D40034P	44			D43100P	16			YA811-YA814
3011Y	20		Y7	WK	0-0	27-62-14	D40034P	46			D43100P	18			YA811-YA814
3011Z	20		Y7	WK	0-0	27-62-14	D40034P	S/R			D43100P	S/R			YA811-YA814
W1034		286A1034			RIGHT WING TRAILING EDGE:STA 525 TO R. WING DISC STA 254R										
0513	20		HL		1-5	27-52-11	D00229	15			GD00136	ST..	E		ALL
1001	20		HP	AA	16-11	28-21-11	D00790	6			D40034J	34			ALL
1001Z				AA	0-0	28-21-11	D00790	S/R			D40034J	S/R			ALL
1002	18		HP	AB	2-8	27-53-21	D11828	8			GD00136	ST..	E	*1	ALL
1002Z				AB	0-0	27-53-21	D11828	S/R			GD00136	ST..	E	*1	ALL
1003	18		HP	AC	8-11	27-53-21	D11828	8			GD00134	ST..	E	*2	ALL
1003Z				AC	0-0	27-53-21	D11826	S/R			GD00134	ST..	E	*2	ALL
2001B	20		Y6	GA	28-0	27-18-11	D03574	1			D40034J	7			ALL
2001R	20		Y6	GA	0-0	27-18-11	D03574	2			D40034J	6			ALL
2001Z				GA	0-0	27-18-11	GD00130	ST..	E		D40034J	S/R			ALL
2002B	20		Y6	GB	13-1	22-11-31	D01697	2			D40034J	2			ALL
2002R	20		Y6	GB	0-0	22-11-31	D01697	1			D40034J	1			ALL
2002Z				GB	0-0	22-11-31	D01697	S/R			D40034J	S/R			ALL
2003B	20		Y6	GC	13-1	22-11-31	D01701	2			D40034J	20			ALL
2003R	20		Y6	GC	0-0	22-11-31	D01701	1			D40034J	19			ALL
2003Z				GC	0-0	22-11-31	D01701	S/R			D40034J	S/R			ALL
2004B	20		Y6	GD	20-10	27-32-11	D00229	5			D40034J	30			ALL
2004R	20		Y6	GD	0-0	27-32-11	D00229	4			D40034J	29			ALL
2004Z				GD	0-0	27-32-11	D00229	S/R			D40034J	S/R			ALL
2005B	20		Y6	GE	20-10	22-11-11	D00229	10			D40034J	25			ALL
2005R	20		Y6	GE	0-0	22-11-11	D00229	9			D40034J	24			ALL
2005Z				GE	0-0	22-11-11	D00229	S/R			D40034J	S/R			ALL
2006B	20		M2	GF	20-10	27-52-11	D00229	2			D40034J	37			ALL
2006R	20		M2	GF	0-0	27-52-11	D00229	1			D40034J	36			ALL
2006ZI				GF	0-0	27-52-11	D00229	S/R			D40034J	38			ALL
2006ZO				GF	0-0	27-52-11	D00229	S/R			D40034J	S/R			ALL
2007B	20		M2	GG	20-10	27-52-11	D00229	3			D40034J	41			ALL
2007R	20		M2	GG	0-0	27-52-11	D00229	11			D40034J	40			ALL
2007ZI				GG	0-0	27-52-11	D00229	S/R			D40034J	39			ALL
2007ZO				GG	0-0	27-52-11	D00229	S/R			D40034J	S/R			ALL
2008B	20		Y6	GI	18-5	27-53-21	D11826	7			D11828	1			ALL
2008R	20		Y6	GI	0-0	27-53-21	D11826	6			D11828	2			ALL
2008Z				GI	0-0	27-53-21	D11826	S/R			D11828	S/R			ALL
2009B	20		Y6	GH	11-10	27-53-21	D11826	1			D40034J	12			ALL

图 5-25　相关导线清单

2. 襟翼位置传感器相关设备件号查询

打开 WDM 的设备清单（Equipment List），查询设备号为 D00229（D229）的设备描述及设备件号。

襟翼位置传感器相关
设备件号的查询

如图 5-26 所示，设备 D229 为插头，该插头件号为 BACC63BP20C16SN。

BOEING				737-700 WIRING DIAGRAM MANUAL		
Equip	Opt	Part Number Part Description	Used On Dwg Vendor	Qty	Diagram Station / WL / BL	Effectivity
D00201	1	BACC66F11F04AE01 PLUG-VHF-2 TRANSCEIVER	81205	1	23-12-21 E001-05/ /	ALL
D00201 T	1	S280W601-116 GROUND BLOCK, ARINC 600	81205	1	23-12-21 */*/*	ALL
D00201A	1	PARTAOFD00201 PLUG-VHF-2 TRANSCEIVER		0	23-12-21 */*/*	ALL
D00201B	1	PARTBOFD00201 PLUG-VHF-2 TRANSCEIVER		0	23-12-21 */*/*	ALL
D00201C	1	PARTCOFD00201 PLUG-VHF-2 TRANSCEIVER		0	23-12-21 */*/*	ALL
D00203	I	BACC63CB16-24SN PLUG-CONDUCTIVE CONNECTOR	81205	1	23-12-11 P8-2/ /	YA811-YA814
D00203 T	1	BACC10KD16 BACKSHELL-CONDUCTIVE	81205	1	23-12-11 P8-2/ /	YA811-YA814
D00209	1	BACC63CB16-24SN PLUG-CONDUCTIVE CONNECTOR	81205	1	23-12-21 P8-3/ /	ALL
D00209 T	1	BACC10KD16 BACKSHELL-CONDUCTIVE	81205	1	23-12-21 P8-3/ /	ALL
D00211	1	BACC45FT16-24S7 PLUG-ATC DUAL CONT	81205	1	34-53-11 P008-29/ /	ALL
D00211 T	1	BACC10JS16 BACKSHELL	81205	1	34-53-11 */*/*	ALL
D00229	1	BACC63BP20C16SN PLUG-CONDUCTIVE	81205	1	27-52-11 ORSS/446/	ALL

图 5-26　相关设备清单

3. 襟翼位置传感器线路故障的排除

打开 WDM 的连线清单（Hook-Up List），查询 D229 的连线信息。

襟翼位置传感器线路
故障的排除

如图 5-27 所示，D229 的 Pin1 ～ Pin3、Pin11、Pin15 所在线路图章节为 27-52-11，Pin4 ～ Pin8 所在线路图章节为 27-32-11，Pin9 ～ Pin10 所在线路图章节为 22-11-11，Pin12 ～ Pin14 所在线路图章节为 22-12-41。

综上所述，按照故障隔离手册中故障隔离程序的检测步骤，结合 WDM 线路图与导线清单、设备清单、连接清单等，可以快速地完成线路故障的排除。

🔊 思考题

1. 你能在 WDM 手册的导线清单中任意找一根导线编号，并说出导线的类型代码、连接设备、所在章节号、有效性等信息吗？

2. 你能在 WDM 手册的设备清单中任意找一设备号，并查询该设备的件号、所有接口的连线、相关的章节号、有效性等信息吗？

BOEING 737-700 WIRING DIAGRAM MANUAL

Equip	Term	Station / WL / BL Type	Bundle	Description Wire No.	GA	CO	Diagram	Effectivity
D00211		P008-29/ /		PLUG-ATC DUAL CONT (continued)				ALL
	5	VP	W2177	0005	20		34-53-11	ALL
	6		W2177	0006	20		34-53-11	ALL
	7		W2177	0007	24		34-53-11	ALL
	8		W2177	0008	20		34-53-11	ALL
	9	VP	W2177	0009	20		34-53-11	ALL
	10	VP	W2177	0010	20		34-53-11	ALL
	11		W2177	0011	24		34-53-11	ALL
	12		W2177	0012	24		34-53-11	ALL
	13	UNUSED						ALL
	14	UNUSED						ALL
	15		W2177	0015	24		34-53-11	ALL
	16		W2177	0016	24		34-53-11	ALL
	17	UNUSED						ALL
	18		W2177	0018	20		33-18-61	ALL
	19	UNUSED						ALL
	20		W2177	0020	24		34-53-11	ALL
	21		W2177	0021	24		33-18-61	ALL
	22		W2177	2001R	24		34-53-11	ALL
	23		W2177	2001B	24		34-53-11	ALL
	24		W2177	0024	24		34-53-11	ALL
	DED		W2177	2001Z			34-53-11	ALL
D00229		ORSS/446/		PLUG-CONDUCTIVE				ALL
	1		W1034	2006R	20		27-52-11	ALL
	2		W1034	2006B	20		27-52-11	ALL
	3		W1034	2007B	20		27-52-11	ALL
	4		W1034	2004R	20		27-32-11	ALL
	5		W1034	2004B	20		27-32-11	ALL
	6		W1034	3004R	20		27-32-11	ALL
	7		W1034	3004B	20		27-32-11	ALL
	8		W1034	3004Y	20		27-32-11	ALL
	9		W1034	2005R	20		22-11-11	ALL
	10		W1034	2005B	20		22-11-11	ALL
	11		W1034	2007R	20		27-52-11	ALL
	12		W1034	3005R	20		22-12-41	ALL
	13		W1034	3005B	20		22-12-41	ALL
	14		W1034	3005Y	20		22-12-41	ALL
	15		W1034	0513	20		27-52-11	ALL
	16	UNUSED						ALL
	S/R		W1034	2004Z			27-32-11	ALL

图 5-27 相关连线清单

工卡实操

本项目配套民航机务维修典型案例—WDM 应用工卡 6 个，专业面向包括 AV 和 ME，难度系数设置了Ⅲ（易）、Ⅱ（中）、Ⅰ（难）三个等级，学习者可按需选用。工卡实操过程中，学习者不仅需要关注修理任务本身，同时需有意识的查询手册中相关任务的保障人身与设备安全的警告（Warning）和注意事项（Caution/Note）。工作中，工作者需要确认已经查询或知晓了相关注意事项和警告，并在工卡上签署确认，才可开始相应任务。本项目配套工卡详情见表 5-9 所示。

表 5-9 项目五配套工卡列表

工卡序号	项目名称	专业方向	难度系数
WDM-01	襟翼位置传感器线路故障的排除	AV/ME	Ⅲ
WDM -02	飞机上设备 M1020 的线路分析	AV/ME	Ⅱ
WDM -03	飞机上跳开关 C00413 的线路分析	AV/ME	Ⅱ
WDM -04	飞机上接线盒 TB32 的线路分析	AV/ME	Ⅱ
WDM -05	飞机 APU 防火测试线路故障的排除	AV/ME	Ⅰ
WDM -06	气象雷达收发组件的故障修理	AV/ME	Ⅰ

➡ 项目评价

"项目5　飞机襟翼位置传感器线路故障的排除"过程考核评价标准					
课程：　　　　　授课教师：　　　　　专业：					
考核内容	考核标准	评价方式			
知识点（25分）： 1. 飞机线路图手册（WDM）的结构和内容。 2. 飞机线路图手册（WDM）的有效性。 3. 飞机线路图手册（WDM）中各组成元件的线路符号和含义。 4. 飞机线路图手册（WDM）中的WL、EQ、HL等清单的查询方法。	□全对（25分）。	中国大学MOOC平台评分（100%）			
法规知识点（20分）： 1. 飞机导线信息查询清单（WL）。 2. 飞机设备信息查询清单（EQ）。 3. 飞机插头连接信息查询清单(HL)。	□制定的施工工卡完全满足施工程序(20分)。每错（漏）一项施工程序扣5分，扣完为止。	教师评分（100%）			
技能点（20分）： 1. 襟翼位置传感器线路图的查询。 2. 襟翼位置传感器导线信息的查询。 3. 襟翼位置传感器跳开关的拆卸。	□襟翼位置传感器跳开关拆卸过程规范，并完成修理任务（20分）。每错一步施工步骤，扣5分，扣完为止。没有完成修理任务可按步骤计分，依次为信息查询10分，拆卸操作10分。	小组自评（10%）	小组互评（20%）	教师评分（50%）	企业导师评分（20%）
工作作风　操作安全（5分）	□正确做好用电安全措施，安全步骤不缺失，计5分，否则计0分。				
工作作风　团结协作（5分）	□与所分配成员相处融洽，积极参与、分工协作，顺利完成任务，得5分，否则计0分。				
工作作风　节约意识（5分）	□合理使用耗材，计5分，出现浪费计0分。				
工作作风　6S管理（5分）	□按6S要求整理操作台、实训场地，计5分，否则计0分。				
安全意识（5分）	□项目报告中，精准表达对安全意识的体会得5分。	教师评分（100%）			
责任使命感（5分）	□项目报告中,对责任使命感体会深刻得5分。				
职业品格（5分）	□项目报告中,对职业品格体会深刻得5分。				

随手笔记

《民用航空飞机维修手册应用》实操工卡

工卡标题 Title	AMM 查询应用 1：飞机发动机燃油渗漏分析 AMM Query & Application 1: Analysis of Fuel Leakage of Aircraft Engine			
机型 A/C Type	B737 6/7/8	难度等级 Difficulty Level		III
机号 REG.NO.	B-****	版本 Revision		R3
组别 Group		组长 Leader	组员 Team	学时 Period	1
参考资料 Reference	飞机维护手册 Aircraft Maintenance Manual			
警告 Warning	在发动机运行时，请勿进入通气排放附近的危险区域。参考 71-00-00/201 对危险区域的定义。如果人员在发动机工作期间进入这些区域可能导致操作人员人身伤害和设备损坏。 DO NOT ENTER THE HAZARD AREAS AROUND THE INLET AND THE EXHAUST DURING ENGINE OPERATION. REFER TO 71-00-00/201 FOR THE DEFINITION OF THE HAZARD AREAS. IF PERSONNEL ENTER THESE AREAS DURING ENGINE OPERATION, INJURY TO PERSONS AND DAMAGE TO EQUIPMENT COULD OCCUR. 使用耗材时要小心。遵守材料制造商的说明和当地法规。 BE CAREFUL WHEN YOU USE CONSUMABLE MATERIALS. OBEY THE MATERIAL MANUFACTURER'S INSTRUCTIONS AND YOUR LOCAL REGULATIONS. 小心不要让飞机燃油进入嘴里或眼睛。 BE CAREFUL NOT TO GET AIRCRAFT FUEL IN YOUR MOUTH OR IN YOUR EYES. 如果你的皮肤或衣服上沾了油，请立即用水冲洗。 IF YOU GET FUEL ON YOUR SKIN OR ON YOUR CLOTHES, FLUSH IT IMMEDIATELY WITH WATER. 如有必要，请就医。 IF NECESSARY, GET MEDICAL HELP. 飞机燃料有毒。 AIRCRAFT FUEL IS POISONOUS. 飞机燃油系统渗漏维护工作 Warning 及思路			

编写人 Author		审核人 Reviewer		批准人 Approver	
编写时间 Author Date		审核时间 Review Date		批准时间 Approve Date	

工具 / 设备 / 材料 Tool/Equipment/Material				工作者 Perf.By	检查者 Insp.By
名称 Name	规格型号 Specification	数量 Quantity	使用情况 Usage		
计算机 Computer	Windows 7	1			
翻译软件 Translation App	有道等 Youdao etc.	1			
阅读器软件 Reader App	PDF 阅读器 PDF Reader App	1			

《民用航空飞机维修手册应用》实操工卡

1. 工作任务 Requirement		工作者 Perf.By	检查者 Insp.By
飞机发动机燃油渗漏分析 Analysis of Fuel Leakage of Aircraft Engine			
2. 工作准备 **Job Set-up**		工作者 Perf.By	检查者 Insp.By
1）准备好相关设备及软件。 1）Prepare equipment and software. （1）Windows 7 及以上系统计算机一台； （1）A computer with Windows 7 system or above; （2）在计算机上安装有道 App 及 PDF 阅读器； （2）Install Youdao App and PDF reader on the computer; （3）在手机上安装有道 App。 （3）Install Youdao App on your mobile phone.			
2）选择有效的技术文件和手册。 2）Select effective technical documents and manuals.			
3）确认注意事项和警告的相关内容。 3）Confirm caution and warning.			
3. 工作步骤 **Procedure**		工作者 Perf.By	检查者 Insp.By
模仿查询飞机 B-**** 燃油系统渗漏标准，并采取相应故障维护措施。 Imitate to Query the Fuel System Leakage Standard of B-****,and Take Corresponding Maintenance Measures. 1. 查找飞机有效性 Find the Effective Aircraft 打开 AMM 手册——EFFECTIVE AIRCRAFT（图略） Turn to EFFECTIVE AIRCRATE of AMM. 确定该手册适用于 B-**** 飞机。 Make sure that the manual is applicable to B-****. 2. 确定发动机燃油渗漏所在章节 Locate the Chapter of Engine Fuel Leakage			

BOEING | CFM56 ENGINES (CFM56-7) | 737-600/700/800/900 AIRCRAFT MAINTENANCE MANUAL

POWER PLANT - ENGINE DRAINS

General

Engine drains prevent fluid contact with hot engine areas. You use engine drains to detect component failures. Engine drains direct these items overboard:

- Oil
- Fuel
- Hydraulic fluid
- Water
- Vapor.

These components drain fluids through the starter air discharge duct in the right fan cowl:

- Strut
- Main oil/fuel heat exchanger
- Hydromechanical unit (HMU)

YUN ALL PRE SB CFM56-7B 73-44
- Burner staging valve (BSV)

YUN ALL
- High pressure turbine active clearance control (HPTACC) valve
- Low pressure turbine active clearance control (LPTACC) valve
- Left and right variable stator vane (VSV) actuators
- Left and right variable bleed valve (VBV) actuators
- Transient bleed valve (TBV).

Fluids drain through a hole in the left fan cowl panel from these components:

- Fuel pump
- Integrated drive generator (IDG)
- Hydraulic pump.

The oil tank drains fluid through a hole in the right fan cowl panel.

See the AMM for more information about allowable leakage limits. (AMM PART II 71-71)

EFFECTIVITY
YUN ALL

71-00-00

D633A101-YUN

Page 12
Jun 10/2003

SDS 中确定发动机燃油渗漏应该查找 71 章发动机。
Engine Fuel Leakage should be searched in Chapter 71-Power Plant according to SDS.

3. 确定发动机燃油渗漏所在节
Find the Subject of Engine Fuel Leakage
打开 71 章目录查找发动机通气排放——检查 / 检验
Click the Contents to go to Chapter 71-Engine Vents and Drains- Inspection/Check

NACELLE WIRING HARNESSES - REMOVAL/ INSTALLATION	71-51-03	401	SHZ ALL
Nacelle Wiring Harnesses Removal TASK 71-51-03-000-801-F00		401	SHZ ALL
Nacelle Wiring Harnesses Installation TASK 71-51-03-400-801-F00		410	SHZ ALL
ENGINE VENTS AND DRAINS - INSPECTION/ CHECK	71-71-00	601	SHZ ALL
Engine Vents and Drains Inspection TASK 71-71-00-200-801-F00		601	SHZ ALL
Drain Lines Inspection (Operational Check) TASK 71-71-00-700-801-F00		622	SHZ ALL
Drain Lines Inspection (General Visual) TASK 71-71-00-200-802-F00		623	SHZ ALL

确定发动机燃油通气排放检查应查询 71-71-00。
The engine fuel vents and drains inspection should be searched in 71-71-00.

班级 Class	工作卡号 Work Card No	AMM-01	共 4 页　第 4 页

4. 查询发动机燃油渗漏标准
Search for Engine Vents and Drains Leakage Limits
打开 71-71-00-200-801-F00 发动机通气排放的检查
Turn to 71-71-00-200-801-F00 Engine Vents and Drains Inspection

  F. Engine Vents and Drains Leakage Limits and Corrective Action
     SUBTASK　71-71-00-790-007-F00
    (1) Fuel drain system
      (a) Fluid: Fuel
      (b) Threshold limit: 180 cc/hr (60 drops per minute)
        1) If the leakage is less than the threshold limit, no maintenance action is necessary.
        2) If the leakage is more than the threshold limit and less than the serviceable limit, you can continue the engine in service for not more than 25 flight cycles before you replace the applicable component.
      (c) Serviceable limit: 270 cc/hr (90 drops per minute)

F. 发动机通气排放渗漏标准和校正动作
SUBTASK 71-71-00-790-007-F00
（1）燃油渗漏系统
液体：燃油
放行标准：180cc/hr（每分钟 60 滴）
1）如果渗漏＜放行标准，不需要维护。
2）如果放行标准＜渗漏＜可服务标准，替换适用部件前还可以飞不超过 25 个循环。
可服务标准：270cc/hr（每分钟 90 滴）

4. 结束工作 Close Out	工作者 Perf.By	检查者 Insp.By
1）保存文件，文件用班级＋学号＋姓名＋AMM-01 的方式命名； 1）Save the file with the name of Class + Student ID + Name + AMM-01; 2）关机； 2）Shut down the computer; 3）6S 管理。 3）Conduct 6S Management.		

随手笔记

《民用航空飞机维修手册应用》实操工卡

班级 Class		工作卡号 Work Card No	AMM-02	共 2 页　第 1 页	

工卡标题 Title	AMM 查询应用 2：飞机后登机门（左）站位查询 AMM Query and Application 2: Query the Position of the Rear Boarding Gate (Left) of the Aircraft				
机型 A/C Type	B737 6/7/8		难度等级 Difficulty Level		II
机号 REG.NO.	B-****		版本 Revision		R3
组别 Group		组长 Leader		组员 Team	学时 Period	1
参考文件 Reference	飞机维护手册 Aircraft Maintenance Manual				
注意事项 Note	机身图为您提供了参考系统，可帮助您找到组件、特征，以及相对于基准平面的主要机身结构开口。基准平面垂直于机身中心线，位于机身前方 130.0 英寸（3.302 米）机鼻处。 The fuselage station diagram gives you a reference system to help you find components, features, and major fuselage structural openings in relation to a datum plane. The datum plane is perpendicular to the fuselage centerline and found 130.0 inches (3.302 meters) forward of the airplane nose.				

编写人 Author		审核人 Reviewer		批准人 Approver	
编写时间 Author Date		审核时间 Review Date		批准时间 Approve Date	

工具 / 设备 / 材料 Tool/Equipment/Material				工作者 Perf.By	检查者 Insp.By
名称 Name	规格型号 Specification	数量 Quantity	使用情况 Usage		
计算机 Computer	Windows 7	1			
翻译软件 Translation App	有道等 Youdao etc.	1			
阅读器软件 Reader App	PDF 阅读器 PDF Reader App	1			

1. 工作任务 Requirement	工作者 Perf.By	检查者 Insp.By
飞机后登机门（左）站位查询 Query the Position of the Rear Boarding Gate (Left) of the Aircraft		

2. 工作准备 Job Set-up	工作者 Perf.By	检查者 Insp.By
1）准备好相关设备及软件。 1）Prepare equipment and software. （1）Windows 7 及以上系统计算机一台； （1）A computer with Windows 7 system or above; （2）在计算机上安装有道 App 及 PDF 阅读器； （2）Install Youdao App and PDF reader on the computer; （3）在手机上安装有道 App。 （3）Install Youdao App on your mobile phone.		

《民用航空飞机维修手册应用》实操工卡

步骤	工作者 Perf.By	检查者 Insp.By
2）选择有效的技术文件和手册。 2）Select effective technical documents and manuals.		
3）确认注意事项和警告的相关内容。 3）Confirm caution and warning.		

3. 工作步骤 Procedure	工作者 Perf.By	检查者 Insp.By
查询飞机 B-**** 后登机门（左）的站位信息 Search for the Station Information of the Rear Gate (Left) of B-**** * 请在手册中截出相应的图片内容说明每一步骤的查询结果 * *Please Use the Corresponding Screenshot in the Manual to Illustrate the Query Results of Each Step. * 查询步骤： Query Steps: 1. 查询飞机有效性 Check Aircraft Effectiveness 飞机 B-**** 的有效性代码（Effective Code）为 _____。 The effective code of B-**** is_____.		
2. 查询飞机后登机门（左）所在章 Search for the Chapter of Rear Gate (Left) 根据《ATA 100 规范》，飞机站位信息应在 _____ 章。 According to ATA100, aircraft station information should be in Chapter _____.		
3. 查询飞机后登机门（左）所在节 Search for the Section of Rear Gate (Left) 飞机机身（Fuselage）上的登机门应查找 _____ 节。 The boarding gate on the fuselage should be searched in Section _____.　　飞机后登机门（左）站位章节查询（包含有效性）		
4. 查询飞机后登机门（左）任务单 Search for the Task of Rear Gate (Left) 根据左、后登机门图上阴影的标注可以确定左登机门的站位任务单号为 _____。 According to the marking of shadow on the left and rear gate map, the station task number of the left gate can be determined as _____.		
5. 确定飞机后登机门（左）站位信息 Search for the Station of Rear Gate (Left) 根据图示可确定飞机后登机门（左）区间段为 ____， 站位（STA）区间从 ____ 到 ____。 According to the diagram, the section of the rear gate (left) of the aircraft can be determined as___, and station (sta) section is from ____ to ____.　　飞机后登机门（左）站位信息（含 Note）		

4. 结束工作 Close Out	工作者 Perf.By	检查者 Insp.By
保存文件，文件用班级 + 学号 + 姓名 +AMM-02 的方式命名； 1）Save the file with the name of Class + Student ID + Name + AMM-02; 2）关机； 2）Shut down the computer; 3）6S 管理。 3）Conduct 6S Management.		

《民用航空飞机维修手册应用》实操工卡

工卡标题 Title	AMM 查询应用 3：飞机在强风下的牵引 AMM query and Application 3: Aircraft Towing in High Winds			
机型 A/C Type	B737 6/7/8		难度等级 Difficulty Level	II
机号 REG.NO.	B-****		版本 Revision	R3
组别 Group		组长 Leader	组员 Team	学时 Period　1
参考资料 Reference	飞机维护手册 Aircraft Maintenance Manual			

警告 Warning & 注意事项 Caution	警告： WARNING: 当你推或拖飞机时，必须确保牵引机头固定绳索与飞机相连。并将头戴式电源线钩连接到电源检修舱的前下角。 WHEN YOU PUSH OR TOW THE AIRPLANE, YOU MUST MAKE SURE THE TOW TRACTOR HEAD SET CORD IS CONNECTED TO THE AIRPLANE. CONNECT THE HEAD SET CORD HOOK TO THE LOWER FORWARD CORNER OF THE ELECTRICAL POWER ACCESS COMPARTMENT. 在飞机移动过程（牵引、后推或滑行）中遵守这些安全预防措施。确保： OBEY THESE SAFETY PRECAUTIONS DURING MOVEMENT OF THE AIRCRAFT (TOWING, PUSHBACK OR TAXIING). MAKE SURE THAT: 飞机路径上没有人员、设备和其他障碍物。 THE PATH OF THE AIRCRAFT IS CLEAR OF PERSONS, EQUIPMENT AND OTHER OBSTACLES. 禁止任何人靠近牵引车、牵引杆、起落架、发动机短舱或飞机机身下方。 NO PERSONS GO NEAR THE TOW TRACTOR, TOWBAR, LANDING GEARS, ENGINE NACELLES OR BELOW THE AIRCRAFT FUSELAGE. 牵引车上只有合格人员，牵引杆上没有人坐或站。 ONLY QUALIFIED PERSONS ARE ON THE TRACTOR AND NO PERSONS SIT OR STAND ON THE TOWBAR. 在飞机完全停止前，任何人不得靠近飞机。 NO PERSONS GO NEAR THE AIRCRAFT BEFORE IT IS FULLY STOPPED. 如果您不遵守这些指示，可能会造成人身伤害或死亡。 THERE IS A RISK OF INJURY OR DEATH IF YOU DO NOT OBEY THESE INSTRUCTIONS. 在牵引 / 滑行操作（包括低速操作）期间，确保飞机上的每个人都坐在一个座位上，并系好安全带。 DURING TOWING/TAXIING OPERATIONS (LOW-SPEED OPERATIONS INCLUDED), MAKE SURE THAT EACH PERSON IN THE AIRCRAFT IS IN A SEAT AND THAT THE SEAT BELT IS FASTENED. 如果没有系好安全带，当飞机突然停飞，就有受伤的危险。 IF THE SEAT BELT IS NOT FASTENED, THERE IS A RISK OF INJURY IF THE AIRCRAFT STOPS SUDDENLY. 注意事项： CAUTION: 如果发动机罩打开，不要在地面上牵引或移动飞机。

飞机在强风下的牵引工作
Warning&Caution

《民用航空飞机维修手册应用》实操工卡

班级 Class		工作卡号 Work Card No	AMM-03	共 3 页　第 2 页

警告 Warning & 注意事项 Caution	DO NOT TOW OR MOVE THE AIRCRAFT ON THE GROUND IF THE ENGINE COWLS ARE OPEN. 机罩打开时，飞机的移动可能会损坏机罩和机舱结构。 MOVEMENT OF THE AIRCRAFT WITH THE COWLS OPEN CAN CAUSE DAMAGE TO THE COWLS AND THE NACELLE STRUCTURE.		
编写人 Author		审核人 Reviewer		批准人 Approver	
编写时间 Author Date		审核时间 Review Date		批准时间 Approve Date	

工具 / 设备 / 材料 Tool/Equipment/Material				工作者 Perf.By	检查者 Insp.By
名称 Name	规格型号 Specification	数量 Quantity	使用情况 Usage		
计算机 Computer	Windows 7	1			
翻译软件 Translation App	有道等 Youdao etc.	1			
阅读器软件 Reader App	PDF 阅读器 PDF Reader App	1			

1.　工作任务 Requirement	工作者 Perf.By	检查者 Insp.By
飞机在强风下的牵引 Aircraft Towing in High Winds		

2.　工作准备 Job Set-up	工作者 Perf.By	检查者 Insp.By
1）准备好相关设备及软件。 1）Prepare equipment and software. （1）Windows 7 及以上系统计算机一台； （1）A computer with Windows 7 system or above; （2）在计算机上安装有道 App 及 PDF 阅读器； （2）Install Youdao App and PDF reader on the computer; （3）在手机上安装有道 App。 （3）Install Youdao App on your mobile phone.		
2）选择有效的技术文件和手册。 2）Select effective technical documents and manuals.		
3）确认注意事项和警告的相关内容。 3）Confirm caution and warning.		

3.　工作步骤 Procedure	工作者 Perf.By	检查者 Insp.By
查询飞机 B-**** 在强风下牵引的条件 Search for the Condition of B-**** Towing in High Wind * 请在手册中截出相应的图片内容说明每一步骤的查询结果 *		

《民用航空飞机维修手册应用》实操工卡

班级 Class	工作卡号 Work Card No	AMM-03	共 3 页　第 3 页	

	工作者 Perf.By	检查者 Insp.By
*Please Use the Corresponding Screenshot in the Manual to Illustrate the Query Results of Each Step. * 查询步骤： Query Steps： 1．查询飞机有效性 Check Aircraft Effectiveness 飞机 B-**** 的有效性代码（Effective Code）为 _____。 The effective code of B-**** is_____.		
2．查询飞机牵引与滑行所在章 Search for the Chapter of Aircraft Towing and Taxiing 根据《ATA 100 规范》，飞机牵引与滑行应在 _____ 章。 According to ATA 100,aircraft towing and taxiing should be in Chapter _____.		
3．查询飞机牵引所在节 Search for the Section of Aircraft Towing 飞机牵引应查找 _____ 节。 Aircraft towing should be find in Section_____.		
4．查询飞机牵引工作任务单（Task） Search for the Aircraft Towing Task 飞机牵引工作任务号为 _____。 The task number of aircraft towing is_____.		
5．查询飞机强风（High Wind）下牵引的子任务单 Search for the Subtask of Aircraft Towing in High Wind 强风下的牵引工作程序任务号为 _____。 The task number of towing in high winds is_____.		
6．确定飞机强风下牵引的条件 Confirm the Conditions of Aircraft Towing in High Winds 可以在大风中牵引飞机，但必须遵守图 _____ 中描述的条件。 It is possible to tow the aircraft in high winds, but it is necessary to follow the conditions Diagram_____ described in. 飞机在强风下的牵引		
4. 结束工作 **Close Out**	工作者 Perf.By	检查者 Insp.By
1）保存文件，文件用班级 + 学号 + 姓名 +AMM-03 的方式命名； 1）Save the file with the name of Class + Student ID + Name + AMM-03； 2）关机； 2）Shut down the computer； 3）6S 管理。 3）Conduct 6S Management.		

随手笔记

《民用航空飞机维修手册应用》实操工卡

班级 Class		工作卡号 Work Card No	AMM-04	共 3 页　第 1 页	

工卡标题 Title	AMM 查询应用 4：飞机液压油箱的勤务 AMM Query & Application 4: Servicing of Aircraft Hydraulic Reservoir				
机型 A/C Type	B737 6/7/8		难度等级 Difficulty Level		II
机号 REG.NO.	B-****		版本 Revision		R3
组别 Group		组长 Leader		组员 Team		学时 Period	1
参考资料 Reference	飞机维护手册 Aircraft Maintenance Manual				

警告 Warning

请勿使液压油流到您身上。液压油 BMS 3-11 可能会导致人身伤害。如果您的皮肤上有液压油，请用水冲洗您的皮肤。如果您的眼睛中有液压油，请用水冲洗您的眼睛，并获得医疗援助。如果您饮食或喝液压油，需获得医疗援助。

DO NOT GET HYDRAULIC FLUID ON YOU. HYDRAULIC FLUID, BMS 3-11 CAN CAUSE INJURY TO PERSONS. IF YOU GET THE HYDRAULIC FLUID ON YOUR SKIN, FLUSH YOUR SKIN WITH WATER. IF YOU GET THE HYDRAULIC FLUID IN YOUR EYES, FLUSH YOUR EYES WITH WATER AND GET MEDICAL AID. IF YOU EAT OR DRINK THE HYDRAULIC FLUID, GET MEDICAL AID.

在对其他液压系统加压之前，确保维修中的液压系统已隔离。

MAKE SURE THAT THE HYDRAULIC SYSTEM(S) IN MAINTENANCE IS(ARE) ISOLATED BEFORE YOU PRESSURIZE THE OTHER HYDRAULIC SYSTEM(S).

不要让液体沾到皮肤上或眼睛里。如果您这样做：

DO NOT GET THE FLUID ON YOUR SKIN OR IN YOUR EYES.
IF YOU DO:
用清水冲洗，
FLUSH IT AWAY WITH CLEAN WATER,
请就医。
GET MEDICAL AID.

飞机液压油箱的勤务
工作 Warning

在给液压系统加压之前，确保控制装置与它们操作的项目的位置一致。液压操作项目的不必要移动可能导致严重伤害和 / 或损坏。

MAKE SURE THAT THE CONTROLS AGREE WITH THE POSITION OF THE ITEMS THEY OPERATE BEFORE YOU PRESSURIZE A HYDRAULIC SYSTEM. UNWANTED MOVEMENT OF HYDRAULICALLY OPERATED ITEMS CAN CAUSE SERIOUS INJURY AND/OR DAMAGE.

仅在相关液压系统加压时操作控制装置。

ONLY OPERATE CONTROLS WHEN THE RELATED HYDRAULIC SYSTEMS ARE PRESSURIZED.

如果在相关液压系统未加压时操作控制装置，则存在以下风险：

IF YOU OPERATE A CONTROL WHEN THE RELATED HYDRAULIC SYSTEM IS NOT PRESSURIZED, THERE IS A RISK THAT:

控件将处于与其操作的项目不一致的位置。

THE CONTROL WILL BE IN A POSITION THAT DOES NOT AGREE WITH THE ITEM(S) IT OPERATES.

《民用航空飞机维修手册应用》实操工卡

警告 Warning	当液压恢复时，液压操作部件将发生不必要的移动，并造成严重伤害和／或损坏。 WHEN HYDRAULIC PRESSURE IS RESTORED, UNWANTED MOVEMENT OF THE HYDRAULICALLY OPERATED ITEM(S) WILL OCCUR AND CAUSE SERIOUS INJURY AND/OR DAMAGE. 使用耗材时要小心。遵守材料制造商的说明和当地法规。 BE CAREFUL WHEN YOU USE CONSUMABLE MATERIALS. OBEY THE MATERIAL MANUFACTURER'S INSTRUCTIONS AND YOUR LOCAL REGULATIONS.

编写人 Author		审核人 Reviewer		批准人 Approver	
编写时间 Author Date		审核时间 Review Date		批准时间 Approve Date	

工具／设备／材料 Tool/Equipment/Material				工作者 Perf.By	检查者 Insp.By
名称 Name	规格型号 Specification	数量 Quantity	使用情况 Usage		
计算机 Computer	Windows 7	1			
翻译软件 Translation App	有道等 Youdao etc.	1			
阅读器软件 Reader App	PDF 阅读器 PDF Reader App	1			

1. 工作任务 Requirement	工作者 Perf.By	检查者 Insp.By
飞机液压油箱的勤务 Servicing of Aircraft Hydraulic Reservoir		

2. 工作准备 Job Set-up	工作者 Perf.By	检查者 Insp.By
1）准备好相关设备及软件。 1）Prepare equipment and software. （1）Windows 7 及以上系统计算机一台； （1）A computer with Windows 7 system or above; （2）在计算机上安装有道 App 及 PDF 阅读器； （2）Install Youdao App and PDF reader on the computer; （3）在手机上安装有道 App。 （3）Install Youdao App on your mobile phone.		
2）选择有效的技术文件和手册。 2）Select effective technical documents and manuals.		
3）确认注意事项和警告的相关内容。 3）Confirm caution and warning.		

《民用航空飞机维修手册应用》实操工卡

3. 工作步骤 Procedure		工作者 Perf.By	检查者 Insp.By
查询飞机 B-**** 液压油箱勤务的工作程序 Search for the Procedure of B-**** Hydraulic Reservoir * 请在手册中截出相应的图片内容说明每一步骤的查询结果 * *Please Use the Corresponding Screenshot in the Manual to Illustrate the Query Results of Each Step. * 查询步骤： Query Steps： 1. 查询飞机有效性 Check Aircraft Effectiveness 飞机 B-**** 的有效性代码（Effective Code）为 _____。 The effective code of B-**** is_____.			
2. 查询飞机勤务所在章 Search for the Chapter of the Aircraft Servicing 根据《ATA 100 规范》，飞机勤务应在 _____ 章。 According to ATA100, the aircraft servicing should be in Chapter _____.			
3. 查询飞机液压油箱所在节 Search for the Section of Aircraft Hydraulic Reservoir 飞机液压油箱勤务应查找 _____ 节。 The aircraft hydraulic reservoir should be searched in Section_____.			
4. 查询飞机液压油箱勤务工作任务单 Search for the Task of Aircraft Hydraulic Reservoir Servicing 飞机液压油箱勤务任务号为 _____。 The task of aircraft hydraulic reservoir servicing is_____.			
5. 确定飞机液压油箱勤务的工作程序 Confirm the Procedure of Aircraft Hydraulic Reservoir Servicing 飞机液压油箱勤务的工作程序在 _____ 段落，低于液压油箱下限值时的工作程序在 _____ 段落。 The working procedure of aircraft hydraulic reservoir servicing is in Paragraph_____, and the working procedure below the lower limit of hydraulic oil tank is in Paragraph _____.	飞机液压油箱的勤务		
4. 结束工作 Close Out		工作者 Perf.By	检查者 Insp.By
1）保存文件，文件用班级 + 学号 + 姓名 +AMM-04 的方式命名； 1）Save the file with the name of Class + Student ID + Name + AMM-04; 2）关机； 2）Shut down the computer; 3）6S 管理。 3）Conduct 6S Management.			

随手笔记

《民用航空飞机维修手册应用》实操工卡

工卡标题 Title	AMM 查询应用 5：飞机空调系统空气混合腔的安装 AMM Query & Application 5: Installation of Mix Chamber in Air Conditioning				
机型 A/C Type	B737 6/7/8		难度等级 Difficulty Level	I	
机号 REG.NO.	B-****		版本 Revision	R3	
组别 Group		组长 Leader		组员 Team		学时 Period	1
参考资料 Reference	飞机维护手册 Aircraft Maintenance Manual				

警告 Warning & 注意事项 Note & 注意事项 Caution	警告： WARNING: 确保没有从主机、APU 或地源向空调系统提供空气。 MAKE SURE THAT AIR IS NOT SUPPLIED TO THE AIR CONDITIONING SYSTEM FROM THE MAIN ENGINE, THE APU OR A GROUND SOURCE. 热压缩空气会对人员造成伤害。 HOT COMPRESSED AIR CAN CAUSE INJURY TO PERSONNEL. 开始工作前，确保气动系统已减压。压缩空气会对人员造成伤害。 MAKE SURE THAT THE PNEUMATIC SYSTEM IS DEPRESSURIZED BEFORE YOU START WORK. PRESSURIZED AIR CAN CAUSE INJURY TO PERSONNEL. 在部件充分清洁之前，不要接触它， DO NOT TOUCH THE COMPONENT UNTIL IT IS SUFFICIENTLY, 冷却以防烧伤。 COOL TO PREVENT BURNS. 注意事项： NOTE: 注意确保排水管没有塌陷。 MAKE SURE YOU DO NOT COLLAPSE THE DRAIN TUBE. 注意事项： CAUTION: 不要把你的重量放在导管上。很容易弄坏它们。 DO NOT PUT YOUR WEIGHT ON THE DUCTS. IT IS VERY EASY TO BREAK THEM. 在空调管道上或附近工作时要非常小心。管道是用薄材料制成的，你很容易把它们弄断。 BE VERY CAREFUL WHEN YOU DO WORK ON OR NEAR THE AIR-CONDITIONING DUCTS. THE DUCTS ARE MADE OF THIN MATERIAL AND YOU CAN EASILY BREAK THEM.

飞机空调系统空气混合腔安装工作
Waring&Caution&Note

编写人 Author		审核人 Reviewer		批准人 Approver	
编写时间 Author Date		审核时间 Review Date		批准时间 Approve Date	

《民用航空飞机维修手册应用》实操工卡

工具 / 设备 / 材料 Tool/Equipment/Material				工作者 Perf.By	检查者 Insp.By
名称 Name	规格型号 Specification	数量 Quantity	使用情况 Usage		
计算机 Computer	Windows 7	1			
翻译软件 Translation App	有道等 Youdao etc.	1			
阅读器软件 Reader App	PDF 阅读器 PDF Reader App	1			

1. 工作任务 Requirement	工作者 Perf.By	检查者 Insp.By
飞机空调系统空气混合腔的安装 Installation of Mix Chamber in Air Conditioning		

2. 工作准备 Job Set-up	工作者 Perf.By	检查者 Insp.By
1）准备好相关设备及软件。 1）Prepare equipment and software. （1）Windows 7 及以上系统计算机一台； （1）A computer with Windows 7 system or above; （2）在计算机上安装有道 App 及 PDF 阅读器； （2）Install Youdao App and PDF reader on the computer; （3）在手机上安装有道 App。 （3）Install Youdao App on your mobile phone.		
2）选择有效的技术文件和手册。 2）Select effective technical documents and manuals.		
3）确认注意事项和警告的相关内容。 3）Confirm caution and warning.		

3. 工作步骤 Procedure	工作者 Perf.By	检查者 Insp.By
查询飞机 B-**** 空调系统空气混合腔的安装程序。 Search for the Mix Chamber Installation Procedure of B-**** Air Conditioning * 请在手册中截出相应的图片内容说明每一步骤的查询结果 * *Please Use the Corresponding Screenshot in the Manual to Illustrate the Query Results of Each Step. * 查询步骤： Query Steps: 1. 查询飞机有效性（Effectivity） Check Aircraft Effectiveness 飞机 B-**** 的有效性代码（Effective Code）为 _____。 The effective code of B-**** is_____.		

《民用航空飞机维修手册应用》实操工卡

	工作者 Perf.By	检查者 Insp.By
2．查询飞机空调系统（Air Conditioning）所在章（Chapter） Search for the Chapter of Air Conditioning 根据《ATA 100 规范》，飞机空调系统应在 _____ 章。 According to ATA100, the air conditioning should be in Chapter_____.		
3．查询飞机空气混合腔（Mix Chamber）所在目（Subject） Search for the Subject of Air Mixing Chamber 飞机空气混合腔应查找 _____ 目（**-**-**）。 The mix chamber should be searched in Subject _____ （**-**-**）		
4．查询飞机空气混合腔安装（Installation）工作任务单（Task） Search for the Task of Air Mixing Chamber Installation 飞机空气混合腔安装任务号为 _____。 The task number of air mixing chamber installation is_____.		
5．确定飞机空气混合腔的安装程序 Confirm the Installation Procedure of Air Mixing Chamber 列出飞机空气混合腔的安装程序。 List the installation procedure of air mixing chamber.　　 飞机空调系统空气 混合腔的安装		
4. 结束工作 Close Out	工作者 Perf.By	检查者 Insp.By
1）保存文件，文件用班级＋学号＋姓名＋AMM-05 的方式命名； 1）Save the file with the name of Class + Student ID + Name + AMM-05; 2）关机； 2）Shut down the computer; 3）6S 管理。 3）Conduct 6S Management.		

随手笔记

....

....

....

....

....

....

《民用航空飞机维修手册应用》实操工卡

工卡标题 Title	AMM 查询应用 6：飞机失速警告系统的测试 AMM Query and Application 6: Testing of Aircraft Stall Warning System

机型 A/C Type	B737 6/7/8	难度等级 Difficulty Level	I
机号 REG.NO.	B-****	版本 Revision	R3

组别 Group		组长 Leader		组员 Team		学时 Period	1

参考资料 Reference	飞机维护手册 Aircraft Maintenance Manual
注意事项 Note	显示屏将显示"存在故障？"。FMC 和 ADIRU 值正常。 The display will show EXISTING FAULTS? The FMC and ADIRU values are correct.

编写人 Author		审核人 Reviewer		批准人 Approver	
编写时间 Author Date		审核时间 Review Date		批准时间 Approve Date	

工具 / 设备 / 材料 Tool/Equipment/Material				工作者 Perf.By	检查者 Insp.By
名称 Name	规格型号 Specification	数量 Quantity	使用情况 Usage		
计算机 Computer	Windows 7	1			
翻译软件 Translation App	有道等 Youdao etc.	1			
阅读器软件 Reader App	PDF 阅读器 PDF Reader App	1			

1. 工作任务 Requirement	工作者 Perf.By	检查者 Insp.By
飞机失速警告系统的测试 Testing of Aircraft Stall Warning System		

2. 工作准备 Job Set-up	工作者 Perf.By	检查者 Insp.By
1）准备好相关设备及软件。 1）Prepare equipment and software. （1）Windows 7 及以上系统计算机一台； （1）A computer with Windows 7 system or above; （2）在计算机上安装有道 App 及 PDF 阅读器； （2）Install Youdao App and PDF reader on the computer; （3）在手机上安装有道 App。 （3）Install Youdao App on your mobile phone.		
2）选择有效的技术文件和手册。 2）Select effective technical documents and manuals.		

《民用航空飞机维修手册应用》实操工卡

	工作者 Perf.By	检查者 Insp.By
3）确认注意事项和警告的相关内容。 3）Confirm caution and warning.		
3. 工作步骤 **Procedure**　　　　　　　　　（工作者 Perf.By）（检查者 Insp.By）		
查询飞机 B-**** 失速警告系统的运行测试程序。 Search for the Testing Procedure B-**** Stall Warning System * 请在手册中截出相应的图片内容说明每一步骤的查询结果 * *Please Use the Corresponding Screenshot in the Manual to Illustrate the Query Results of Each Step. * 查询步骤： Query Steps: 1. 查询飞机有效性（Effectivity） Check Aircraft Effectiveness 飞机 B-**** 的有效性代码（Effective Code）为 _____。 The effective code of B-**** is _____.		
2. 查询飞机失速警告系统（Stall Warning System）所在章（Chapter） Search for the Chapter of Stall Warning System 根据《ATA 100 规范》，飞机飞行控制系统应在 _____ 章。 According to ATA100, the flight controls should be in Chapter_____.		
3. 查询飞机失速警告系统所在节（Section） Search for the Section of Stall Warning System 飞机失速警告系统应查找 _____ 节（**-**）。 The stall warning system should be searched in Section_____（**-**-**）.		
4. 查询飞机失速警告系统的运行测试（Operational Test）工作任务单（Task） Search for the Operational Test Task of Stall Warning System 飞机失速警告系统的运行测试任务号为 _____。 The operational test task number of stall warning system is_____.		
5. 确定飞机失速警告系统的运行测试程序 Confirm the Operational Test Procedure of Stall Warning System 列出飞机失速警告系统的运行测试程序。 List the operational test procedure of stall warning system.　　飞机失速警告系统的测试		
4. 结束工作 **Close Out**　　　　　　　　　（工作者 Perf.By）（检查者 Insp.By）		
1）保存文件，文件用班级 + 学号 + 姓名 +AMM-06 的方式命名； 1）Save the file with the name of Class + Student ID + Name + AMM-06; 2）关机； 2）Shut down the computer; 3）6S 管理。 3）Conduct 6S Management.		

《民用航空飞机维修手册应用》实操工卡

工卡标题 Title	AMM 查询应用 7：飞机客舱座椅的安装 AMM Query and Application 7: Installation of Aircraft Passenger Seats		
机型 A/C Type	B737 6/7/8	难度等级 Difficulty Level	I
机号 REG.NO.	B-****	版本 Revision	R3

组别 Group		组长 Leader		组员 Team		学时 Period	1

参考资料 Reference	飞机维护手册 Aircraft Maintenance Manual

警告 Warning & 注意事项 Caution	警告： WARNING: 拆卸或安装此设备时要小心。此设备很重［超过 12kg（26.5 lb）］，可能导致人身伤害和／或损坏。 BE CAREFUL WHEN YOU REMOVE OR INSTALL THIS EQUIPMENT. THIS EQUIPMENT IS HEAVY (MORE THAN 12kg (26.5 lb)) AND CAN CAUSE INJURY AND/OR DAMAGE. 注意事项： CAUTION: 请勿过分拧紧螺丝。如果过分拧紧螺丝，可能会损坏螺丝的固定特性。 DO NOT TIGHTEN THE SCREW TOO MUCH. YOU CAN CAUSE DAMAGE TO THE CAPTURE FEATURE OF THE SCREW IF YOU TIGHTEN THE SCREW TOO MUCH.	飞机客舱座椅的安装工作 Warning&Caution

编写人 Author		审核人 Reviewer		批准人 Approver	
编写时间 Author Date		审核时间 Review Date		批准时间 Approve Date	

工具 / 设备 / 材料 Tool/Equipment/Material				工作者 Perf.By	检查者 Insp.By
名称 Name	规格型号 Specification	数量 Quantity	使用情况 Usage		
计算机 Computer	Windows 7	1			
翻译软件 Translation App	有道等 Youdao etc.	1			
阅读器软件 Reader App	PDF 阅读器 PDF Reader App	1			

1. 工作任务 Requirement	工作者 Perf.By	检查者 Insp.By
飞机客舱座椅的安装 Installation of Aircraft Passenger Seats		

《民用航空飞机维修手册应用》实操工卡

2. 工作准备 Job Set-up	工作者 Perf.By	检查者 Insp.By
1）准备好相关设备及软件。 1）Prepare equipment and software. （1）Windows 7 及以上系统计算机一台； （1）A computer with Windows 7 system or above; （2）在计算机上安装有道 App 及 PDF 阅读器； （2）Install Youdao App and PDF reader on the computer; （3）在手机上安装有道 App。 （3）Install Youdao App on your mobile phone.		
2）选择有效的技术文件和手册。 2）Select effective technical documents and manuals.		
3）确认注意事项和警告的相关内容。 3）Confirm caution and warning.		

3. 工作步骤 Procedure	工作者 Perf.By	检查者 Insp.By
查询飞机客舱座椅的安装程序 Search for the Installation Procedure of Aircraft Passenger Seats * 请在手册中截出相应的图片内容说明每一步骤的查询结果 * *Please Use the Corresponding Screenshot in the Manual to Illustrate the Query Results of Each Step. * 查询步骤： Query Steps: 1. 查询飞机有效性（Effectivity） Check Aircraft Effectiveness 飞机 B-**** 的有效性代码（Effective Code）为 _____。 The effective code of B-**** is_____.		
2. 查询飞机上设备 / 设施（Equipment/Furnishings）所在章（Chapter） Search for the Chapter of Aircraft Equipment/Furnishings 根据《ATA 100 规范》，飞机设备 / 设施应在 _____ 章。 According to ATA100, the aircraft equipment/Furnishings should be in Chapter _____.		
3. 查询飞机客舱座椅（Passenger Seats）所在节（Section） Search for the Section of Aircraft Passenger Seats 飞机客舱座椅应查找 _____ 节（**-**）。 The aircraft passenger seats should be searched in Section _____（**-**-**）.		
4. 查询飞机客舱座椅安装（installation）的工作任务单（Task） Search for the Task of Aircraft Passenger Seats Installation 飞机客舱座椅安装的任务号为 _____。 The task number of aircraft passenger seats installation is_____.		
5. 确定飞机客舱座椅安装的程序 Confirm the Installation Procedure of Aircraft Passenger Seats 列出飞机客舱座椅安装的程序。 List the installation procedure of aircraft passenger seats.		

《民用航空飞机维修手册应用》实操工卡

4. 结束工作 Close Out	工作者 Perf.By	检查者 Insp.By
1）保存文件，文件用班级＋学号＋姓名＋AMM-07 的方式命名； 1）Save the file with the name of Class + Student ID + Name + AMM-07; 2）关机； 2）Shut down the computer; 3）6S 管理。 3）Conduct 6S Management.		

随手笔记

《民用航空飞机维修手册应用》实操工卡

工卡标题 Title	IPC 查询应用 1：飞机前厕所烟雾探测组件的安装 IPC Query & Application 1: Installation of Smoke Detector Assembly in the Lavatory in front of the Aircraft			
机型 A/C Type	B737 6/7/8	难度等级 Difficulty Level		III
机号 REG.NO.	B-****	版本 Revision		R3
组别 Group		组长 Leader	组员 Team	学时 Period	1
参考资料 Reference	飞机维护手册 Aircraft Maintenance Manual 图解零部件目录手册 Illustrated Parts Catalog			
注意事项 Note	安装时请勿拧紧安装螺钉。天花板支架只能松散地安装在烟雾探测器上。 DO NOT TIGHTEN THE MOUNTING SCREWS AT THIS TIME. THE CEILING BRACKET SHOULD ONLY BE ATTACHED LOOSELY TO THE SMOKE DETECTOR.			
编写人 Author		审核人 Reviewer		批准人 Approver	
编写时间 Author Date		审核时间 Review Date		批准时间 Approve Date	

工具 / 设备 / 材料 Tool/Equipment/Material				工作者 Perf.By	检查者 Insp.By
名称 Name	规格型号 Specification	数量 Quantity	使用情况 Usage		
计算机 Computer	Windows 7	1			
翻译软件 Translation App	有道等 Youdao etc.	1			
阅读器软件 Reader App	PDF 阅读器 PDF Reader App	1			
1. 工作任务 Requirement				工作者 Perf.By	检查者 Insp.By
飞机前厕所烟雾探测组件的安装 Installation of Smoke Detector Assembly in the Lavatory in front of the Aircraft					
2. 工作准备 Job Set-up				工作者 Perf.By	检查者 Insp.By

1）准备好相关设备及软件。

1) Prepare equipment and software.

（1）Windows 7 及以上系统计算机一台；

(1) A computer with Windows 7 system or above;

（2）在计算机上安装有道 App 及 PDF 阅读器；

(2) Install Youdao App and PDF reader on the computer;

（3）在手机上安装有道 App。

(3) Install Youdao App on your mobile phone.

《民用航空飞机维修手册应用》实操工卡

	工作者 Perf.By	检查者 Insp.By
2) 选择有效的技术文件和手册。 2) Select effective technical documents and manuals.		
3) 确认注意事项和警告的相关内容。 3) Confirm caution and warning.		

3. 工作步骤 Procedure	工作者 Perf.By	检查者 Insp.By
模仿并查询飞机 B-**** 前厕所烟雾探测组件的件号、供应商 Imitate to Search for the Part Number and supplier of B-**** Smoke Detector Assembly in the Lavatory in front of the Aircraft 查询步骤： Query Steps: 1. 查询飞机的有效性（图略） Check Aircraft Effectiveness		
2. 确定烟雾探测应查找防火系统（FIRE PROTECTION） Confirm the Smoke Detector Should be searched in the Chapter of Fire Protection		
3. 确定厕所烟雾探测器（LAVALORY SMOKE DETECTOR）应查询 SECTION 26-14 Confirm the Lavalory Smoke Detector Should be Searched in Section 26-14 		

查询可得如图示两个烟雾探测器，项目为前厕所烟雾探测组件，因此确定为第二幅图站位为 STA178 ～ STA360 之间的项目号为 55 的组件。

Two smoke detectors are shown in the figures. The item is the smoke detector assembly in the lavatory in front of the aircraft, therefore, it is determined as the component with item No. 55 between STA178 and STA360 in the second figure.

4. 在部件目录（Parts Catalog）中查询前厕所烟雾探测组件的相关信息
Search for the Related information of Smoke Detector Assembly in the Lavatory in front of the Aircraft in the Parts Catalog

FIG ITEM	PART NUMBER	1234567　　NOMENCLATURE	EFFECT FROM TO	UNITS PER ASSY
8		STRUCTURE INSTL-LAV A (LAVATORY SMOKE DETECTORS ONLY)		
- 1	25-45-16-025RF	STRUCTURE INSTL-LAV A (LAVATORY SMOKE DETECTORS ONLY) SUPPLIER CODE: VU3384 FOR NHA/OTHER SYS DET SEE: 25-40-01-56A	001001	RF
- 1	25-45-16-033RF	STRUCTURE INSTL-LAV A (LAVATORY SMOKE DETECTORS ONLY) SUPPLIER CODE: VU3384 FOR NHA/OTHER SYS DET SEE: 25-40-01-56A	002999	RF
50	C22102-101-309	.SERVICE UNIT ASSY-PASS. SUPPLIER CODE: VU3384		RF
55	C22102-294-001	...DETECTOR ASSY-SMOKE SUPPLIER CODE: VU3384 ELECTRICAL EQUIP NUMBER: MZ0109 ALTERED FROM PART NUMBER: PU90-499R3 VS5065		1
60	MS35206-229	...SCREW-		2

《民用航空飞机维修手册应用》实操工卡

5. 确定飞机前厕所烟雾探测组件的安装信息件号为 C22102-294-001；项目号为 55；供应商号码为 VU3384 Confirm that the Installation Information Part Number of the Smoke Detector Assembly in the Lavatory in front of the Aircraft is C22102-294-001; The Item Number is 55; The Supplier Number is VU3384 飞机前厕所烟雾探测组件的安装		
4. 结束工作 **Close Out**	工作者 Perf.By	检查者 Insp.By
1）保存文件，文件用班级 + 学号 + 姓名 +IPC-01 的方式命名； 1）Save the file with the name of Class + Student ID + Name + IPC-01; 2）关机； 2）Shut down the computer; 3）6S 管理。 3）Conduct 6S Management.		

随手笔记

《民用航空飞机维修手册应用》实操工卡

工卡标题 Title	IPC 查询应用 2：飞机上 284A2841-1 件的更换 IPC Query and Application 2: Replacement of 284A2841-1 on the Aircraft			
机型 A/C Type	B737 6/7/8	难度等级 Difficulty Level		II
机号 REG.NO.	B-****	版本 Revision		R3

组别 Group		组长 Leader		组员 Team		学时 Period	1

参考资料 Reference	飞机维护手册 Aircraft Maintenance Manual 图解零部件目录手册 Illustrated Parts Catalog
注意事项 Note	卸下面板时，确保泡沫密封件已连接到飞机结构上。 MAKE SURE THE FOAM SEALS ARE ATTACHED TO THE AIRPLANE STRUCTURE WHEN YOU REMOVE THE PANEL.

编写人 Author		审核人 Reviewer		批准人 Approver	
编写时间 Author Date		审核时间 Review Date		批准时间 Approve Date	

工具 / 设备 / 材料 Tool/Equipment/Material				工作者 Perf.By	检查者 Insp.By
名称 Name	规格型号 Specification	数量 Quantity	使用情况 Usage		
计算机 Computer	Windows 7	1			
翻译软件 Translation App	有道等 Youdao etc.	1			
阅读器软件 Reader App	PDF 阅读器 PDF Reader App	1			

1. 工作任务 Requirement	工作者 Perf.By	检查者 Insp.By
飞机上 284A2841-1 件的更换 Replacement of 284A2841-1 on the Aircraft		

2. 工作准备 Job Set-up	工作者 Perf.By	检查者 Insp.By

1）准备好相关设备及软件。
1）Prepare equipment and software.
（1）Windows 7 及以上系统计算机一台；
（1）A computer with Windows 7 system or above;
（2）在计算机上安装有道 App 及 PDF 阅读器；
（2）Install Youdao App and PDF reader on the computer;

《民用航空飞机维修手册应用》实操工卡

	工作者 Perf.By	检查者 Insp.By
（3）在手机上安装有道 App。 （3）Install Youdao App on your mobile phone.		
2）选择有效的技术文件和手册。 2）Select effective technical documents and manuals.		
3）确认注意事项和警告的相关内容。 3）Confirm caution and warning.		
3. 工作步骤 **Procedure**	**工作者** **Perf.By**	**检查者** **Insp.By**
查询飞机 B-**** 上件号为 284A2841-1 的零部件的名称、安装位置等信息。 Search for the Name, Installation Location and Other Information of the Part No. 284A2841-1 on B-****. * 请在手册中截出相应的图片内容说明每一步骤的查询结果 * *Please Use the Corresponding Screenshot in the Manual to Illustrate the Query Results of Each Step. * 查询步骤： Query Steps: 1. 查询飞机有效性（Effectivity） Check Aircraft Effectiveness 飞机 B-**** 的有效性代码（Effective Code）为 _____。 The effective code of B-**** is_____.		
2. 查询件号为 284A2841-1 的零部件的章 - 节 - 单元 - 图号 - 项目号（Chapter-Section-UNIT-Figure-Item） Search for the Chapter-Section-Unit-Figure-Item of the Part No. 284A2841-1 通过零部件件号（Parts Number）数字索引表，查询件号为 284A2841-1 的零部件的章 - 节 - 单元为 _____，图号为 _____，项目号为 _____。 Through the parts number alpha-numerical index table, the chapter-section-unit number of the part number 284A2841-1 is_____, the Figure is _____, and the Item is_____.		
3. 确定飞机上件号为 284A2841-1 的零部件的名称 Confirm the Name of the Part Number 284A2841-1 on the Aircraft 查询部件目录（Parts Catalog）可知，飞机上件号为 284A2841-1 零部件为 _____ According to the parts catalog, the part number 284A2841-1 on the aircraft is_____.		
4. 确定待查件 284A2841-1 的零部件装配图 Confirm the Assembly Drawing of the Part 284A2841-1 to be Checked 根据待查零部件项目号，可得该件的装配图。 According to the item number of the part to be checked, the assembly drawing of the part can be obtained. 飞机上 284A2841-1 件的更换		

班级 Class	工作卡号 Work Card No	IPC-02	共 3 页　第 3 页

4. 结束工作 Close Out	工作者 Perf.By	检查者 Insp.By
1）保存文件，文件用班级 + 学号 + 姓名 +IPC-02 的方式命名； 1）Save the file with the name of Class + Student ID + Name + IPC-02; 2）关机； 2）Shut down the computer; 3）6S 管理。 3）Conduct 6S Management.		

随手笔记

《民用航空飞机维修手册应用》实操工卡

工卡标题 Title	IPC 查询应用 3：飞机上 AS120G120A000 件的更换 IPC Query and Application 3: Replacement of AS120G120A000 on the Aircraft			
机型 A/C Type	B737 6/7/8	难度等级 Difficulty Level	II	
机号 REG.NO.	B-****	版本 Revision	R3	
组别 Group		组长 Leader	组员 Team	学时 Period	1

参考资料 Reference	飞机维护手册 Aircraft Maintenance Manual 图解零部件目录手册 Illustrated Parts Catalog
警告 Warning	消除气动系统的压力。气动系统中的压力可能导致人身伤害和设备损坏。 REMOVE THE PRESSURE FROM THE PNEUMATIC SYSTEM. PRESSURE IN THE PNEUMATIC SYSTEM CAN CAUSE INJURY TO PERSONS AND DAMAGE TO EQUIPMENT

编写人 Author		审核人 Reviewer		批准人 Approver	
编写时间 Author Date		审核时间 Review Date		批准时间 Approve Date	

工具 / 设备 / 材料 Tool/Equipment/Material				工作者 Perf.By	检查者 Insp.By
名称 Name	规格型号 Specification	数量 Quantity	使用情况 Usage		
计算机 Computer	Windows 7	1			
翻译软件 Translation App	有道等 Youdao etc.	1			
阅读器软件 Reader App	PDF 阅读器 PDF Reader App	1			
1. 工作任务 **Requirement**				工作者 Perf.By	检查者 Insp.By
飞机上 AS120G120A000 件的更换 Replacement of AS120G120A000 on the Aircraft					
2. 工作准备 **Job Set-up**				工作者 Perf.By	检查者 Insp.By

1）准备好相关设备及软件。
1）Prepare equipment and software.
（1）Windows 7 及以上系统计算机一台；
（1）A computer with Windows 7 system or above；
（2）在计算机上安装有道 App 及 PDF 阅读器；
（2）Install Youdao App and PDF reader on the computer；
（3）在手机上安装有道 App。
（3）Install Youdao App on your mobile phone.

《民用航空飞机维修手册应用》实操工卡

2）选择有效的技术文件和手册。 2）Select effective technical documents and manuals.		
3）确认注意事项和警告的相关内容。 3）Confirm caution and warning.		

3. 工作步骤 Procedure	工作者 Perf.By	检查者 Insp.By
查询飞机 B-**** 上件号为 AS120G120A000 的零部件的名称、安装位置等信息 Search for the Name, Installation Location and Other Information of the Part No. AS120G120A000 on B-**** * 请在手册中截出相应的图片内容说明每一步骤的查询结果 * *Please Use the Corresponding Screenshot in the Manual to Illustrate the Query Results of Each Step. * 查询步骤: Query Steps: 1. 查询飞机有效性(Effectivity) Check Aircraft Effectiveness 飞机 B-**** 的有效性代码(Effective Code)为 _____。 The effective code of B-**** is_____.		
2. 查询件号为 AS120G120A000 的零部件的章 - 节 - 单元 - 图号 - 项目号(Chapter- Section- Unit-Figure-Item) Search for the Chapter-Section-Unit-Figure-Item of the Part No. AS120G120A000 通过零部件件号(Parts Number)字母索引表,查询件号为 AS120G120A000 的零部件的章 - 节 - 单元为 _____, 图号为 _____, 项目号为 _____。 Through the parts number alpha-numerical index table, the chapter-section-unit number of the part number AS120G120A000 is_____, the Figure is _____, and the Item is_____.		
3. 确定飞机上件号为 AS120G120A000 的零部件的名称 Confirm the Name of the Part Number AS120G120A000 on the Aircraft 查询部件目录(Parts Catalog)可知,飞机上件号为 AS120G120A000 的零部件为 _____。 According to the parts catalog, the part number AS120G120A000 on the aircraft is_____.		
4. 确定待查件 AS120G120A000 的零部件装配图 Confirm the Assembly Drawing of the Part AS120G120A000 to be Checked 根据待查零部件项目号,可得该件的装配图。 According to the item number of the part to be checked, the assembly drawing of the part can be obtained.　飞机上 AS120G120A000 件的更换		

4. 结束工作 Close Out	工作者 Perf.By	检查者 Insp.By
1）保存文件,文件用班级 + 学号 + 姓名 +IPC-03 的方式命名; 1）Save the file with the name of Class + Student ID + Name + IPC-03; 2）关机; 2）Shut down the computer; 3）6S 管理。 3）Conduct 6S Management.		

《民用航空飞机维修手册应用》实操工卡

<table>
<tbody>
<tr>
<td>工卡标题
Title</td>
<td colspan="5">IPC 查询应用 4：飞机前轮舱灯组件的更换
IPC Inquiry & Application 4: Replacement of the Light Assembly in the Aircraft Front Wheel Well</td>
</tr>
<tr>
<td>机型
A/C Type</td>
<td colspan="2">B737 6/7/8</td>
<td>难度等级
Difficulty Level</td>
<td colspan="2">I</td>
</tr>
<tr>
<td>机号
REG.NO.</td>
<td colspan="2">B-****</td>
<td>版本
Revision</td>
<td colspan="2">R3</td>
</tr>
<tr>
<td>组别
Group</td>
<td></td>
<td>组长
Leader</td>
<td>组员
Team</td>
<td>学时
Period</td>
<td>1</td>
</tr>
<tr>
<td>参考资料
Reference</td>
<td colspan="5">飞机维护手册
Aircraft Maintenance Manual
图解零部件目录手册
Illustrated Parts Catalog</td>
</tr>
<tr>
<td>注意事项
Note</td>
<td colspan="5">飞机前轮舱灯组件的更换需严格遵循 AMM TASK 33-32-00-960-801 轮舱灯更换的工作程序。
THE REPLACEMENT OF THE FRONT WHEEL WELL LIGHT ASSEMBLY OF THE AIRCRAFT MUST STRICTLY FOLLOW THE WORKING PROCEDURES OF THE AMM TASK 33-32-00-960-801 WHEEL WELL LIGHT REPLACEMENT WORK PROCEDURE.</td>
</tr>
<tr>
<td>编写人
Author</td>
<td></td>
<td colspan="2">审核人
Reviewer</td>
<td colspan="2">批准人
Approver</td>
</tr>
<tr>
<td>编写时间
Author Date</td>
<td></td>
<td colspan="2">审核时间
Review Date</td>
<td colspan="2">批准时间
Approve Date</td>
</tr>
</tbody>
</table>

工具 / 设备 / 材料 Tool/Equipment/Material				工作者 Perf.By	检查者 Insp.By
名称 Name	规格型号 Specification	数量 Quantity	使用情况 Usage		
计算机 Computer	Windows 7	1			
翻译软件 Translation App	有道等 Youdao etc.	1			
阅读器软件 Reader App	PDF 阅读器 PDF Reader App	1			

1. 工作任务 Requirement	工作者 Perf.By	检查者 Insp.By
飞机前轮舱灯组件的更换 Replacement of the Light Assembly in the Aircraft Front Wheel Well		

2. 工作准备 Job Set-up	工作者 Perf.By	检查者 Insp.By
1）准备好相关设备及软件。 1）Prepare equipment and software. （1）Windows 7 及以上系统计算机一台； （1）A computer with Windows 7 system or above; （2）在计算机上安装有道 App 及 PDF 阅读器； （2）Install Youdao App and PDF reader on the computer;		

内容	工作者 Perf.By	检查者 Insp.By
（3）在手机上安装有道 App。 （3）Install Youdao App on your mobile phone.		
2）选择有效的技术文件和手册。 2）Select effective technical documents and manuals.		
3）确认注意事项和警告的相关内容。 3）Confirm caution and warning.		

3. 工作步骤 Procedure	工作者 Perf.By	检查者 Insp.By
查询飞机 B-**** 前轮舱灯组件的件号、供应商。 Search for the Part Number and supplier of B-**** Front Wheel Well Light Assembly of the Aircraft * 请在手册中截出相应的图片内容说明每一步骤的查询结果 * *Please Use the Corresponding Screenshot in the Manual to Illustrate the Query Results of Each Step. * 查询步骤： Query Steps: 1. 查询飞机有效性（Effectivity） Check Aircraft Effectiveness 飞机 B-**** 的有效性代码（Effective Code）为 _____。 The effective code of B-**** is_____.		
2. 查询飞机灯光系统所在章（Chapter） Search for the Chapter of Aircraft Lighting System 根据《ATA 100 规范》，飞机前轮舱灯应在 _____ 章。 According to ATA100, the aircraft wheel well lights should be in Chapter_____.		
3. 查询飞机前轮舱灯（Front Wheel Well Lights）所在节（Section） Search for the Section of Aircraft Front Wheel Well Lights 飞机前轮舱灯应查找 _____ 节。 Aircraft Front Wheel Well Lights should be searched in Section_____.		
4. 查询飞机前轮舱灯组件项目号（Item） Search for the Item of Aircraft Front Wheel Well Light Assembly 飞机前轮舱灯组件项目号为 _____。 The Item of aircraft front wheel well light assembly is_____.		
5. 确定飞机前轮舱灯组件相关信息 Confirm the Relevant Information of Aircraft Front Wheel Well Light Assembly 根据部件目录（Parts Catalog）可知，飞机前轮舱灯组件件号为 _____，飞机前轮舱灯（Lamp）件号为 _____。 According to the parts catalog,the part number of Aircraft front wheel well light assembly is_____, the part number of lamp is_____. 飞机前轮舱灯组件的更换		

《民用航空飞机维修手册应用》实操工卡

4．结束工作 Close Out	工作者 Perf.By	检查者 Insp.By
1）保存文件，文件用班级＋学号＋姓名＋IPC-04 的方式命名； 1）Save the file with the name of Class + Student ID + Name + IPC-04; 2）关机； 2）Shut down the computer; 3）6S 管理。 3）Conduct 6S Management.		

随手笔记

随手笔记

《民用航空飞机维修手册应用》实操工卡

工卡标题 Title	IPC 查询应用 5：飞机尾白航行灯的安装 IPC Inquiry & Application 5: Installation of White Position Lights on the Aircraft Tail			
机型 A/C Type	B737 6/7/8		难度等级 Difficulty Level	I
机号 REG.NO.	B-****		版本 Revision	R3
组别 Group		组长 Leader	组员 Team	学时 Period	1
参考资料 Reference	飞机维护手册 Aircraft Maintenance Manual 图解零部件目录手册 Illustrated Parts Catalog			

警告 Warning & 注意事项 Note	警告： WARNING: 在给飞机供电之前，确保所有维修中的电路都已隔离。不需要的电力可能是危险的。 MAKE SURE THAT ALL THE CIRCUITS IN MAINTENANCE ARE ISOLATED BEFORE YOU SUPPLY ELECTRICAL POWER TO THE AIRCRAFT. UNWANTED ELECTRICAL POWER CAN BE DANGEROUS. 在以下区域或附近开始作业前，请将安全装置和警告标志放置到位： PUT THE SAFETY DEVICES AND THE WARNING NOTICES IN POSITION BEFORE YOU START A TASK ON OR NEAR: 飞行控制； THE FLIGHT CONTROLS; 飞行操纵面； THE FLIGHT CONTROL SURFACES; 起落架和相关舱门； THE LANDING GEAR AND THE RELATED DOORS; 移动的组件。 COMPONENTS THAT MOVE. 部件的移动可能导致人员伤亡和 / 或设备损坏。 MOVEMENT OF COMPONENTS CAN KILL OR CAUSE INJURY TO PERSONS AND/ OR CAN CAUSE DAMAGE TO THE EQUIPMENT. 开始工作前，必须系上安全带并将其连接到检修平台上。没系安全带可能导致摔倒致死或受伤。 BEFORE YOU START WORK, YOU MUST PUT ON A SAFETY HARNESS AND ATTACH IT TO THE ACCESS PLATFORM. WITHOUT A SAFETY HARNESS, YOU CAN FALL. THIS CAN KILL YOU OR CAUSE YOU INJURY. 操作后 5 分钟内不要触摸闪光灯。闪光灯仍然是热的。 DO NOT TOUCH THE STROBE LIGHTS FOR AT LEAST 5 MINUTES AFTER OPERATION. THE STROBE LIGHT WILL STILL BE HOT. 在灯光下工作时，戴上眼镜和防护服。灯被加压，可能发生爆炸。 WEAR GLASSES AND PROTECTIVE CLOTHING WHEN YOU DO WORK ON THE LIGHT. THE LAMP IS PRESSURIZED AND AN EXPLOSION CAN OCCUR. 注意事项： NOTE: 每个照明灯组件中都有两个灯。当需要更换一盏灯时，建议同时更换第二个灯。 THERE ARE TWO LAMPS IN EACH LIGHT ASSEMBLY. WHEN IT IS NECESSARY TO REPLACE ONE LAMP, IT IS RECOMMENDED THAT THE SECOND LAMP BE REPLACED AT THE SAME TIME.

飞机尾白航行灯的安装
工作 Waring&Note

《民用航空飞机维修手册应用》实操工卡

编写人 Author		审核人 Reviewer		批准人 Approver	
编写时间 Author Date		审核时间 Review Date		批准时间 Approve Date	

工具 / 设备 / 材料 Tool/Equipment/Material				工作者 Perf.By	检查者 Insp.By
名称 Name	规格型号 Specification	数量 Quantity	使用情况 Usage		
计算机 Computer	Windows 7	1			
翻译软件 Translation App	有道等 Youdao etc.	1			
阅读器软件 Reader App	PDF 阅读器 PDF Reader App	1			

1. 工作任务 Requirement	工作者 Perf.By	检查者 Insp.By
飞机尾白航行灯的安装 Installation of White Position Lights on the Aircraft Tail		

2. 工作准备 Job Set-up	工作者 Perf.By	检查者 Insp.By
1）准备好相关设备及软件。 1）Prepare equipment and software. （1）Windows 7 及以上系统计算机一台； （1）A computer with Windows 7 system or above; （2）在计算机上安装有道 App 及 PDF 阅读器； （2）Install Youdao App and PDF reader on the computer; （3）在手机上安装有道 App。 （3）Install Youdao App on your mobile phone.		
2）选择有效的技术文件和手册。 2）Select effective technical documents and manuals.		
3）确认注意事项和警告的相关内容。 3）Confirm caution and warning.		

3. 工作步骤 Procedure	工作者 Perf.By	检查者 Insp.By
查询飞机 B-**** 尾白航行灯组的件号、供应商。 Search for the Part Number and supplier of B-**** White Position Lights on Aircraft Tail * 请在手册中截出相应的图片内容说明每一步骤的查询结果 * *Please Use the Corresponding Screenshot in the Manual to Illustrate the Query Results of Each Step. * 查询步骤： Query Steps:		

《民用航空飞机维修手册应用》实操工卡

班级 Class	工作卡号 Work Card No	IPC-05	共 3 页　第 3 页

1. 查询飞机有效性（Effectivity） Check Aircraft Effectiveness 飞机 B-**** 的有效性代码（Effective Code）为 _____ 。 The effective code of B-**** is_____.		
2. 查询飞机灯光系统所在章（Chapter） Search for the Chapter of Aircraft Lighting System 根据《ATA 100 规范》，飞机尾白航行灯应在 _____ 章。 According to ATA100, the white position lights on aircraft tail should be in Chapter_____.		
3. 查询飞机航行灯（Position Lights）所在节（Section） Search for the Section of Aircraft Position Lights 飞机尾白航行灯应查找 _____ 节。 The white position lights on aircraft tail should be in Section_____.		
4. 查询飞机尾白航行灯组（After Position）项目号（Item） Search for the Item of Aircraft After Position Lights 飞机尾白航行灯项目号为 _____ 。 The Item of the white position lights on the aircraft tail is_____.		
5. 确定飞机尾白航行灯组的相关信息 Confirm the Relevant Information of Aircraft After Position Lights 根据部件目录（Parts Catalog）可知，飞机尾白航行灯组件号为 _____ ，供应商代码为 _____ 。 According to the parts catalog,the part number of aircraft after position lights is_____, and the supplier code is_____. 飞机尾白航行灯的安装		
4. 结束工作 Close Out	工作者 Perf.By	检查者 Insp.By
1）保存文件，文件用班级 + 学号 + 姓名 +IPC-05 的方式命名； 1）Save the file with the name of Class + Student ID + Name + IPC-05; 2）关机； 2）Shut down the computer; 3）6S 管理。 3）Conduct 6S Management.		

随手笔记

《民用航空飞机维修手册应用》实操工卡

工卡标题 Title	IPC 查询应用 6：飞机垂尾放电刷的更换 IPC Inquiry & Application 6: Replacement of the Static Discharger on the Aircraft Vertical Tail			
机型 A/C Type	B737 6/7/8	难度等级 Difficulty Level		I
机号 REG.NO.	B-****	版本 Revision		R3
组别 Group		组长 Leader		组员 Team		学时 Period	1
参考资料 Reference	飞机维护手册 Aircraft Maintenance Manual 图解零部件目录手册 Illustrated Parts Catalog			
警告 Warning	确保从液压系统中消除压力。确保未提供液压和电力。如果存在液压或提供了液压 / 电力，则可以移动飞行控制表面。这可能会造成人身伤害或设备损坏。 MAKE SURE PRESSURE IS REMOVED FROM HYDRAULIC SYSTEMS. MAKE SURE HYDRAULIC POWER AND ELECTRICAL POWER ARE NOT SUPPLIED. IF HYDRAULIC PRESSURE IS PRESENT OR HYDRAULIC/ELECTRICAL POWER IS SUPPLIED, THE FLIGHT CONTROL SURFACE CAN MOVE. THIS CAN CAUSE INJURY TO PERSONS OR DAMAGE TO EQUIPMENT. 在以下区域或附近开始作业前，请将安全装置和警告标志放置到位： PUT THE SAFETY DEVICES AND THE WARNING NOTICES IN POSITION BEFORE YOU START A TASK ON OR NEAR: 飞行控制； THE FLIGHT CONTROLS 飞行操纵面； THE FLIGHT CONTROL SURFACES 起落架和相关舱门； THE LANDING GEAR AND THE RELATED DOORS 移动的组件。 COMPONENTS THAT MOVE. 部件的移动可能导致人员伤亡和 / 或设备损坏。 MOVEMENT OF COMPONENTS CAN KILL OR CAUSE INJURY TO PERSONS AND/OR CAN CAUSE DAMAGE TO THE EQUIPMENT. 确保地面安全锁在起落架上就位。这将防止起落架不必要的移动，以及可能造成的人员伤害和飞机和 / 或设备损坏。 MAKE SURE THAT THE GROUND SAFETY-LOCKS ARE IN POSITION ON THE LANDING GEAR. THIS WILL PREVENT UNWANTED MOVEMENT OF THE LANDING GEAR, AND THUS POSSIBLE INJURY TO PERSONS AND DAMAGE TO THE AIRCRAFT AND/OR EQUIPMENT.			
编写人 Author		审核人 Reviewer		批准人 Approver	
编写时间 Author Date		审核时间 Review Date		批准时间 Approve Date	

《民用航空飞机维修手册应用》实操工卡

班级 Class	工作卡号 Work Card No	IPC-06	共 3 页　第 2 页

工具 / 设备 / 材料 Tool/Equipment/Material				工作者 Perf.By	检查者 Insp.By
名称 Name	规格型号 Specification	数量 Quantity	使用情况 Usage		
计算机 Computer	Windows 7	1			
翻译软件 Translation App	有道等 Youdao etc.	1			
阅读器软件 Reader App	PDF 阅读器 PDF Reader App	1			

1. 工作任务 Requirement	工作者 Perf.By	检查者 Insp.By
飞机垂尾放电刷的更换 Replacement of the Static Discharger on the Aircraft Vertical Tail		

2. 工作准备 Job Set-up	工作者 Perf.By	检查者 Insp.By
1）准备好相关设备及软件。 1）Prepare equipment and software. （1）Windows 7 及以上系统计算机一台； （1）A computer with Windows 7 system or above; （2）在计算机上安装有道 App 及 PDF 阅读器； （2）Install Youdao App and PDF reader on the computer; （3）在手机上安装有道 App。 （3）Install Youdao App on your mobile phone.		
2）选择有效的技术文件和手册。 2）Select effective technical documents and manuals.		
3）确认注意事项和警告的相关内容。 3）Confirm caution and warning.		

3. 工作步骤 Procedure	工作者 Perf.By	检查者 Insp.By
查询飞机 B-**** 垂尾放电刷的件号、供应商等信息 Search for the Part Number and supplier of B-**** Vertical Tail Static Discharger * 请在手册中截出相应的图片内容说明每一步骤的查询结果 * *Please Use the Corresponding Screenshot in the Manual to Illustrate the Query Results of Each Step. * 查询步骤： Query Steps: 1. 查询飞机有效性（Effectivity） Check Aircraft Effectiveness 飞机 B-**** 的有效性代码（Effective Code）为 _____ The effective code of B-**** is_____		

《民用航空飞机维修手册应用》实操工卡

班级 Class	工作卡号 Work Card No	IPC-06	共 3 页　第 3 页

2．查询飞机通信系统（Communications System）所在章（Chapter） Search for the Chapter of Aircraft Communications System 根据《ATA 100 规范》，飞机通讯系统应在_____章。 According to ATA100, the Aircraft Communications System should be in Chapter_____.		
3．查询飞机静电放电（Static Discharger）所在节（Section） Search for the Section of Aircraft Static Discharger 飞机垂尾放电刷应查找_____节。 Aircraft vertical tail static discharger should be searched in Section_____.		
4．查询飞机垂尾放电刷项目号（Item） Search for the Item of Aircraft Vertical Tail Static Discharger 飞机垂尾放电刷项目号为_____（不唯一）。 The Item of aircraft vertical tail static discharger is_____ (Not unique).		
5．确定飞机垂尾放电刷的相关信息 Confirm the Relevant Information of Aircraft Vertical Tail Static Discharger 根据部件目录（Parts Catalog）可知，飞机垂尾放电刷件号为_____，供应商代码为_____。 According to the parts catalog,the part number of aircraft vertical tail static discharger is_____, the supplier code is_____.　　　　飞机垂尾放电刷的更换		

4．结束工作 Close Out	工作者 Perf.By	检查者 Insp.By
1）保存文件，文件用班级＋学号＋姓名＋IPC-06 的方式命名； 1）Save the file with the name of Class + Student ID + Name + IPC-06; 2）关机； 2）Shut down the computer; 3）6S 管理。 3）Conduct 6S Management.		

随手笔记

13-3

随手笔记

《民用航空飞机维修手册应用》实操工卡

工卡标题 Title	FIM 查询应用 1：APU 无法启动的故障隔离 FIM Query & Application 1: Fault Isolation that APU Cannot Start		
机型 A/C Type	B737 6/7/8	难度等级 Difficulty Level	III
机号 REG.NO.	B-****	版本 Revision	R3

组别 Group		组长 Leader		组员 Team		学时 Period	1
参考资料 Reference		故障隔离手册 Fault Isolation Manual 飞机维护手册 Aircraft Maintenance Manual					
注意事项 Note		您必须执行"使用外部电源进行交换机检查"和"APU 电源"测试，以确保 APU 可以使用外部电源运行。 YOU MUST DO THE SWITCH CHECK WITH EXTERNAL POWER AND APU POWER TEST TO MAKE SURE THE APU CAN OPERATE ON EXTERNAL POWER.					
编写人 Author		审核人 Reviewer			批准人 Approver		
编写时间 Author Date		审核时间 Review Date			批准时间 Approve Date		

工具 / 设备 / 材料 Tool/Equipment/Material				工作者 Perf.By	检查者 Insp.By
名称 Name	规格型号 Specification	数量 Quantity	使用情况 Usage		
计算机 Computer	Windows 7	1			
翻译软件 Translation App	有道等 Youdao etc.	1			
阅读器软件 Reader App	PDF 阅读器 PDF Reader App	1			

1. 工作任务 Requirement	工作者 Perf.By	检查者 Insp.By
APU 无法启动的故障隔离 Fault Isolation that APU Cannot Start		

2. 工作准备 Job Set-up	工作者 Perf.By	检查者 Insp.By

1）准备好相关设备及软件。
1）Prepare equipment and software.
（1）Windows 7 及以上系统计算机一台；
（1）A computer with Windows 7 system or above;
（2）在计算机上安装有道 App 及 PDF 阅读器；
（2）Install Youdao App and PDF reader on the computer;
（3）在手机上安装有道 App。
（3）Install Youdao App on your mobile phone.

2）选择有效的技术文件和手册。
2）Select effective technical documents and manuals.

班级 Class	工作卡号 Work Card No	FIM-01	共 4 页　第 2 页

3）确认注意事项和警告的相关内容。
3）Confirm caution and warning.

3. 工作步骤 Procedure	工作者 Perf.By	检查者 Insp.By
模仿查询飞机 B-****APU 无法启动、FAULT 灯亮，且自检信息代码为 49-41010 故障的可能原因 Imitate to query the possible cause of the APU start failure, FAULT Light being on, and the self-test fault code is 49-41010. * 请在手册中截出相应的图片内容说明每一步骤的查询结果 * *Please Use the Corresponding Screenshot in the Manual to Illustrate the Query Results of Each Step. * 查询步骤: Query Steps: 1. 查询飞机的有效性（图略） Check Aircraft Effectiveness		
2. 可在故障清单中查询 "APU 无法启动，FAULT 灯亮" 的故障描述 Search for the Fault Description of "APU Start Failure, FAULT Light being on" in the Fault List APU 　• air inlet door does not fully open or fully close . . . .　491 010 00　49-60 TASK 801 　• backfire . . . .　492 150 00　49-55 TASK 801 　• does not start, FAULT light on . . . .　492 100 00　49-60 TASK 801 　• does not start, MAINT light on . . . .　492 110 00　49-60 TASK 801 　• does not start, no APU indication lights on . . . .　492 130 00　49-15 TASK 805 　• does not start, OVER SPEED light on . . . .　492 120 00　49-60 TASK 801 　• fire false alarm . . . .　261 060 00　26-10 TASK 801 　• fuel valve does not fully open or fully close . . . .　493 010 00　49-60 TASK 801 　• oil consumption high . . . .　492 140 00　49-95 TASK 801 　• Surges during main engine start . . . .　492 160 00　49-55 TASK 808 　• vibration is excessive . . . .　491 020 00　49-15 TASK 807		
3. 根据与故障描述对应的 FIM 工卡号，在系统中找到 49-60 TASK 801，即可找到 APU 自检程序 According to the FIM Task Number Corresponding to the Fault Description, Find the APU BITE Procedure From 49-60 TASK 801 		

班级 Class	工作卡号 Work Card No	FIM-01	共 4 页　第 3 页

4. 在维护信息索引中查询自检维护代码为 49-41010 的故障隔离工卡
Query the BITE Maintenance Code of 49-41010 Fault Isolation Task in the Maintenance Information Index

LRU/SYSTEM	MAINTENANCE MESSAGE	GO TO FIM TASK
APU	49-31159 FCU RESOLVER CIRCUIT SHOWS SHORT TO GROUND	49-32 TASK 801
APU	49-31160 FCU RESOLVER CIRCUIT SHOWS OPEN	49-32 TASK 802
APU	49-31161 FCU RESOLVER CIRCUIT SHOWS SHORT	49-32 TASK 803
APU	49-31162 FCU RESOLVER CIRCUIT SHOWS BAD	49-32 TASK 804
APU	49-31163 FUEL CONTROL UNIT SHOWS SOLENOID FAILED ON	49-31 TASK 801
APU	49-31164 FUEL CONTROL UNIT SHOWS SOLENOID OPEN CIRCUIT	49-31 TASK 802
APU	49-31166 FUEL CONTROL UNIT SHOWS SOLENOID HIGH CURRENT	49-31 TASK 803
APU	49-31167 FUEL CONTROL UNIT SHOWS FUEL TEMP SENSOR SHORT	49-31 TASK 804
APU	49-31168 FUEL CONTROL UNIT TEMP SHOWS OUT OF RANGE HIGH	49-31 TASK 805
APU	49-31169 FUEL CONTROL UNIT TEMP SHOWS OUT OF RANGE LOW	49-31 TASK 806
APU	49-31170 FUEL CONTROL TORQ MOTOR SHOWS LOW SIDE GROUNDED	49-31 TASK 807
APU	49-31171 FUEL CONTROL UNIT FLOW DISAGREES WITH COMMAND	49-31 TASK 808
APU	49-31172 FUEL CONTROL UNIT TORQUE MOTOR SHOWS OPEN CIRCUIT	49-31 TASK 809
APU	49-31173 FUEL CONTROL UNIT TORQUE MOTOR SHOWS SHORT	49-31 TASK 810
APU	49-31175 FCU RESOLVER CIRCUIT SHOWS BAD	49-32 TASK 805
APU	49-31214 APU FUEL VALVE SHOWS NOT OPEN	49-33 TASK 801
APU	49-31294 FLOW DIVIDER SOLENOID SHOWS HIGH CURRENT	49-33 TASK 802
APU	49-31295 FLOW DIVIDER SOLENOID SHOWS OPEN CIRCUIT	49-33 TASK 803
APU	49-41010 NO ACCELERATION SHUTDOWN	49-40 TASK 801

5. 打开 49-40 TASK 801 工卡，并仔细阅读工卡找到故障可能的原因
Open the 49-40 TASK 801, And Read the Task Carefully to Find the Possible Cause of the Fault

801. **No Acceleration Shutdown - Fault Isolation**

  A. Description

    (1) This task is for this maintenance message:

      (a) 49-41010 NO ACCELERATION SHUTDOWN

    (2) There is no or low acceleration for 12.5 seconds during an APU start cycle.and an APU lite-off occurred.

    (3) The APU FAULT light on the P5 forward overhead panel will show for this maintenance message.

  B. Possible Causes

    (1) Operational problem with the AC generation system

    (2) Operational problem with the DC generation system

    (3) Blockage of unwanted materials in the air inlet duct and compressor inlet plenum for the APU

    (4) Fuel leakage from the APU

    (5) Fuel supply flow problem

    (6) Fuel control unit, YAAM002

    (7) Fuel flow divider, YAAV004

    (8) Linkage problem with the inlet guide vane (IGV) actuator, YAAV002

    (9) Inlet temperature sensor, YAAT002

班级 Class	工作卡号 Work Card No	FIM-01	共 4 页　第 4 页

(10) Inlet pressure sensor (P2), YAAT003 (11) Start converter unit, M1710 (12) Start power unit, M1850 (13) Electronic control unit, M1709 (14) EGT thermocouple 1, YAAT009 (15) EGT thermocouple 2, YAAT010 (16) Starter-generator, YAAG013 (17) APU. 该故障可能的原因在 B 段落，最可能的原因是第（1）项，最不可能的原因是第（17）项，依次排列。 The possible causes of this fault are in paragraph B, the most probable cause is item (1), and the least likely cause is item (17), in order.	APU 无法启动 的故障隔离	
4. 结束工作 **Close Out**	**工作者** **Perf.By**	**检查者** **Insp.By**
1）保存文件，文件用班级＋学号＋姓名＋FIM-01 的方式命名； 1）Save the file with the name of Class + Student ID + Name +FIM-01; 2）关机； 2）Shut down the computer; 3）6S 管理。 3）Conduct 6S Management.		

随手笔记

《民用航空飞机维修手册应用》实操工卡

班级 Class	工作卡号 Work Card No	FIM-02	共 2 页　第 1 页

工卡标题 Title	FIM 查询应用 2：飞机上故障代码为 21002000 的故障隔离 FIM Query & Application 2: Fault Isolation on the Aircraft with the Fault Code 21002000		
机型 A/C Type	B737 6/7/8	难度等级 Difficulty Level	II
机号 REG.NO.	B-****	版本 Revision	R3

组别 Group		组长 Leader		组员 Team		学时 Period	1

参考资料 Reference	故障隔离手册 Fault Isolation Manual 飞机维护手册 Aircraft Maintenance Manual
警告 Warning	触摸过滤器之前，请穿上个人防护设备。过滤器从空气中清除小颗粒（烟、灰尘、棉绒、纤维、花粉）和传染性材料（细菌、病毒、真菌、真菌），会对人造成疾病和伤害。 PUT ON THE PERSONAL PROTECTIVE EQUIPMENT BEFORE YOU TOUCH THE FILTER. THE FILTER REMOVES SMALL PARTICLES (SMOKE, DUST, LINT, FIBERS, POLLEN) AND INFECTIOUS MATERIALS (BACTERIA, VIRUSES, MOLD SPORES, FUNGI) FROM THE AIR WHICH CAN CAUSE ILLNESSES AND INJURIES TO PERSONS.

编写人 Author		审核人 Reviewer		批准人 Approver	
编写时间 Author Date		审核时间 Review Date		批准时间 Approve Date	

工具 / 设备 / 材料 Tool/Equipment/Material				工作者 Perf.By	检查者 Insp.By
名称 Name	规格型号 Specification	数量 Quantity	使用情况 Usage		
计算机 Computer	Windows 7	1			
翻译软件 Translation App	有道等 Youdao etc.	1			
阅读器软件 Reader App	PDF 阅读器 PDF Reader App	1			
1. 工作任务 **Requirement**				工作者 Perf.By	检查者 Insp.By
飞机上故障代码为 21002000 的故障隔离 Fault Isolation on the Aircraft with the Fault Code 21002000					
2. 工作准备 **Job Set-up**				工作者 Perf.By	检查者 Insp.By

1）准备好相关设备及软件。
1）Prepare equipment and software.
（1）Windows 7 及以上系统计算机一台；
（1）A computer with Windows 7 system or above;
（2）在计算机上安装有道 App 及 PDF 阅读器；
（2）Install Youdao App and PDF reader on the computer;

《民用航空飞机维修手册应用》实操工卡

班级 Class	工作卡号 Work Card No	FIM-02	共 2 页 第 2 页

	工作者 Perf.By	检查者 Insp.By
（3）在手机上安装有道 App。 （3）Install Youdao App on your mobile phone.		
2）选择有效的技术文件和手册。 2）Select effective technical documents and manuals.		
3）确认注意事项和警告的相关内容。 3）Confirm caution and warning.		
3. 工作步骤 **Procedure**	工作者 Perf.By	检查者 Insp.By
FRM 给出飞机 B-**** 故障代码为 21002000，请查询该代码的故障隔离程序、故障最可能的原因等。 FRM gives the aircraft B-**** fault code 21002000, please check the fault isolation procedure of this code, the most likely cause of the fault, etc. * 请在手册中截出相应的图片内容说明每一步骤的查询结果 * *Please Use the Corresponding Screenshot in the Manual to Illustrate the Query Results of Each Step. * 查询步骤： Query Steps: 1. 查询飞机有效性（Effectivity） Check Aircraft Effectiveness 飞机 B-**** 的有效性代码（Effective Code）为 _____。 The effective code of B-**** is_____.		
2. 查询故障代码为 21002000 的故障隔离工卡 Query the Fault Isolation Work Card with the Fault Code 21002000 通过可观察故障清单（Observed Fault List），故障代码为 21002000 的故障隔离工卡号为 _____。 Through the observed fault list, the FIM task number for the fault cade of 21002000 is _____.		
3. 确定故障代码为 21002000 的故障隔离程序、故障原因等信息 Confirm the Fault Isolation Procedure and the Cause of the Fault with the Fault Code 21002000 打开并阅读故障隔离工卡可知，该故障隔离程序在 _____ 段落，故障最可能的原因是 _____。 Open and read the fault isolation task. It can be seen that the fault isolation procedure is in the paragraph of _____, and the most possible cause of the fault is _____. 飞机上故障代码为 21002000 的故障隔离		
4. 结束工作 **Close Out**	工作者 Perf.By	检查者 Insp.By
1）保存文件，文件用班级＋学号＋姓名+FIM-02 的方式命名； 1）Save the file with the name of Class + Student ID + Name +FIM-02; 2）关机； 2）Shut down the computer; 3）6S 管理。 3）Conduct 6S Management.		

《民用航空飞机维修手册应用》实操工卡

班级 Class		工作卡号 Work Card No	FIM-03	共 2 页　第 1 页	

工卡标题 Title	FIM 查询应用 3：飞机上故障代码为 28201101 的故障隔离 FIM Query & Application 3: Fault Isolation on the Aircraft with the Fault Code 28201101				
机型 A/C Type	B737 6/7/8		难度等级 Difficulty Level		II
机号 REG.NO.	B-****		版本 Revision		R3
组别 Group		组长 Leader		组员 Team	学时 Period　1
参考资料 Reference	故障隔离手册 Fault Isolation Manual 飞机维护手册 Aircraft Maintenance Manual				
注意事项 Note	如果按住开关超过二十（20）秒，则测试将自动停止，并且指示器返回到指示模式。 IF THE SWITCH IS HELD FOR MORE THAN TWENTY (20) SECONDS, THE TEST STOPS AUTOMATICALLY AND THE INDICATORS GO BACK TO INDICATING MODE.				
编写人 Author		审核人 Reviewer		批准人 Approver	
编写时间 Author Date		审核时间 Review Date		批准时间 Approve Date	

工具 / 设备 / 材料 Tool/Equipment/Material				工作者 Perf.By	检查者 Insp.By
名称 Name	规格型号 Specification	数量 Quantity	使用情况 Usage		
计算机 Computer	Windows 7	1			
翻译软件 Translation App	有道等 Youdao etc.	1			
阅读器软件 Reader App	PDF 阅读器 PDF Reader App	1			

1. 工作任务 Requirement	工作者 Perf.By	检查者 Insp.By
飞机上故障代码为 28201101 的故障隔离 Fault Isolation on the Aircraft with the Fault Code 28201101		

2. 工作准备 Job Set-up	工作者 Perf.By	检查者 Insp.By

1）准备好相关设备及软件。
1）Prepare equipment and software.
（1）Windows 7 及以上系统计算机一台；
（1）A computer with Windows 7 system or above;
（2）在计算机上安装有道 App 及 PDF 阅读器；
（2）Install Youdao App and PDF reader on the computer;
（3）在手机上安装有道 App。
（3）Install Youdao App on your mobile phone.

《民用航空飞机维修手册应用》实操工卡

班级 Class	工作卡号 Work Card No	FIM-03	共 2 页　第 2 页

内容	工作者 Perf.By	检查者 Insp.By
2）选择有效的技术文件和手册。 2）Select effective technical documents and manuals.		
3）确认注意事项和警告的相关内容。 3）Confirm caution and warning.		

3.　工作步骤 Procedure	工作者 Perf.By	检查者 Insp.By
FRM 给出飞机 B-**** 故障代码为 28201101，请查询该代码的故障隔离程序、故障最可能的原因等。 FRM gives the B-**** fault code 28201101. Please query the fault isolation procedure of this code, and the most likely cause of the fault, etc. * 请在手册中截出相应的图片内容说明每一步骤的查询结果 * *Please Use the Corresponding Screenshot in the Manual to Illustrate the Query Results of Each Step. * 查询步骤： Query Steps: 1．查询飞机有效性（Effectivity） Check Aircraft Effectiveness 飞机 B-**** 的有效性代码（Effective Code）为 _____。 The effective code of B-**** is_____.		
2．查询故障代码为 28201101 的故障隔离工卡 Query the fault isolation work card with the fault code 28201101. 通过可观察故障清单（Observed Fault List），查询故障代码为 28201101 的故障隔离工卡号为 _____。 Through the Observed Fault List, The FIM task number for the fault code of 28201101 is _____.		
3．确定故障代码为 28201101 的故障隔离程序、故障原因等信息 Confirm the Fault Isolation Procedure and the Cause of the Fault with the Fault Code 28201101 打开并阅读故障隔离工卡可知，该故障隔离程序在 _____ 段落，故障最可能的原因是 _____。 飞机上故障代码为 28201101 的故障隔离 Open and read the fault isolation task. It can be seen that the fault isolation procedure is in the paragraph of _____, and the most possible cause of the fault is _____.		

4.　结束工作 Close Out	工作者 Perf.By	检查者 Insp.By
1）保存文件，文件用班级＋学号＋姓名＋FIM-03 的方式命名； 1）Save the file with the name of Class + Student ID + Name +FIM-03; 2）关机； 2）Shut down the computer; 3）6S 管理。 3）Conduct 6S Management.		

《民用航空飞机维修手册应用》实操工卡

工卡标题 Title	FIM 查询应用 4：飞机发动机滑油超温（琥珀色）故障隔离 FIM Query and Application 4: Fault Isolation of Engine Oil Overtemperature (Amber)

机型 A/C Type	B737 6/7/8	难度等级 Difficulty Level	I
机号 REG.NO.	B-****	版本 Revision	R3

组别 Group		组长 Leader		组员 Team		学时 Period	1

参考资料 Reference	故障隔离手册 Fault Isolation Manual 飞机维护手册 Aircraft Maintenance Manual
注意事项 Note	如果在维修或更换组件后需要发动机运行作为安装后的测试，可以同时执行此步骤。 IF ENGINE OPERATION IS NECESSARY AS A POST-INSTALLATION TEST AFTER YOU REPAIR OR REPLACE A COMPONENT, YOU CAN DO THIS STEP AT THE SAME TIME.

编写人 Author		审核人 Reviewer		批准人 Approver	
编写时间 Author Date		审核时间 Review Date		批准时间 Approve Date	

工具 / 设备 / 材料 Tool/Equipment/Material				工作者 Perf.By	检查者 Insp.By
名称 Name	规格型号 Specification	数量 Quantity	使用情况 Usage		
计算机 Computer	Windows 7	1			
翻译软件 Translation App	有道等 Youdao etc.	1			
阅读器软件 Reader App	PDF 阅读器 PDF Reader App	1			

1. 工作任务 Requirement	工作者 Perf.By	检查者 Insp.By
飞机发动机滑油超温（琥珀色）故障隔离 Fault Isolation of Engine Oil Overtemperature (Amber)		

2. 工作准备 Job Set-up	工作者 Perf.By	检查者 Insp.By

1）准备好相关设备及软件。
1) Prepare equipment and software.
（1）Windows 7 及以上系统计算机一台；
（1）A computer with Windows 7 system or above;
（2）在计算机上安装有道 App 及 PDF 阅读器；
（2）Install Youdao App and PDF reader on the computer;
（3）在手机上安装有道 App。
（3）Install Youdao App on your mobile phone.

2）选择有效的技术文件和手册。
2) Select effective technical documents and manuals.

3）确认注意事项和警告的相关内容。 3）Confirm caution and warning.		
3. 工作步骤 **Procedure**	工作者 Perf.By	检查者 Insp.By
查询飞机 B-**** 发动机滑油超温（琥珀色）故障隔离程序 Inquire About the Fault Isolation Procedure of the B-**** Engine Oil Over Temperature (Amber) * 请在手册中截出相应的图片内容说明每一步骤的查询结果 * *Please Use the Corresponding Screenshot in the Manual to Illustrate the Query Results of Each Step. * 查询步骤: Query Steps: 1. 查询飞机有效性（Effectivity） Check Aircraft Effectiveness 飞机 B-**** 的有效性代码（Effective Code）为 _____。 The effective code of B-**** is_____.		
2. 查询飞机发动机滑油（Engine Oil）温度指示（Temperature indication）超高（琥珀色）的故障隔离工卡（TASK） Query the FIM Task of the Super High (Amber) Temperature Indication of the Aircraft Engine Oil 通过可观察故障清单（Observed Fault List）字母索引（Alphabetical Index），飞机发动机滑油超温（琥珀色）（Amber）的故障隔离工卡号为 _____。 Through the Observed Fault List alphabetical index, the FIM task work card number of the aircraft engine oil overtemperature (amber) is _____.		
3. 确定发动机滑油超温（琥珀色）故障原因 Determine the Cause of the Engine Oil Overtemperature (Amber) Failure 飞机发动机滑油超温（琥珀色）故障最可能的原因是 _____。 The most likely cause of the aircraft engine oil overtemperature (amber) fault is _____.		
4. 确定发动机滑油超温（琥珀色）故障相关跳开关信息 Determine the Trip Switch Information Related to the Engine Oil Over-Temperature (Amber) Fault 通过阅读飞机发动机滑油超温（琥珀色）故障隔离工卡可知故障相关跳开关信息。其中，发动机为 ____ 号，相关跳开关 C01391 在面板 _____ 的 _____ 行、_____ 列。 By reading the fault isolation work card of the aircraft engine oil overtemperature (amber), the fault-related trip switch information can be known. Among them, the engine number is _____, and the relevant breaker C01391 is in the_____ row, _____ column of the panel _____.		
5. 确定发动机滑油超温（琥珀色）故障隔离程序 Determine the Engine Oil Over-Temperature (Amber) Fault Isolation Procedure 在发动机滑油超温（琥珀色）故障隔离程序中，找出油箱油位的检查程序。 In the engine oil overtemperature (amber) fault isolation procedure, find out the inspection procedure of the oil level for the oil tank. 飞机发动机滑油超温（琥珀色）故障隔离		

《民用航空飞机维修手册应用》实操工卡

班级 Class	工作卡号 Work Card No	FIM-04	共 3 页　第 3 页

4．结束工作 Close Out	工作者 Perf.By	检查者 Insp.By
1）保存文件，文件用班级＋学号＋姓名＋FIM-04 的方式命名； 1）Save the filc with the name of Class + Student ID + Name +FIM-04; 2）关机； 2）Shut down the computer; 3）6S 管理。 3）Conduct 6S Management.		

随手笔记

《民用航空飞机维修手册应用》实操工卡

班级 Class		工作卡号 Work Card No	FIM-05	共 3 页　第 1 页	

工卡标题 Title	FIM 查询应用 5：飞机空调区域灯不工作的故障隔离 FIM Query & Application 5: Fault Isolation of Aircraft Air-Conditioning Area Light Failure				
机型 A/C Type	B737 6/7/8		难度等级 Difficulty Level	I	
机号 REG.NO.	B-****		版本 Revision	R3	
组别 Group		组长 Leader		组员 Team	学时 Period　1
参考资料 Reference	故障隔离手册 Fault Isolation Manual				
注意事项 Caution	更换面板 P5-17 上的客舱温度控制选择器。松开夹子时，请按住选择器。松开夹子后，选择器可能会掉落。如果摔落，选择器可能会损坏。 REPLACE THE CONTROL CABIN TEMPERATURE SELECTOR IN THE P5-17 PANEL. HOLD THE SELECTOR WHEN YOU LOOSEN THE CLAMP. THE SELECTOR CAN FALL WHEN THE CLAMP IS LOOSE. THE SELECTOR CAN BE DAMAGED IF IT FALLS.				
编写人 Author		审核人 Reviewer		批准人 Approver	
编写时间 Author Date		审核时间 Review Date		批准时间 Approve Date	

工具 / 设备 / 材料 Tool/Equipment/Material				工作者 Perf.By	检查者 Insp.By
名称 Name	规格型号 Specification	数量 Quantity	使用情况 Usage		
计算机 Computer	Windows 7	1			
翻译软件 Translation App	有道等 Youdao etc.	1			
阅读器软件 Reader App	PDF 阅读器 PDF Reader App	1			

1. 工作任务 Requirement	工作者 Perf.By	检查者 Insp.By
飞机空调区域灯不工作的故障隔离 Fault Isolation of Aircraft Air-Conditioning Area Light Failure		

2. 工作准备 Job Set-up	工作者 Perf.By	检查者 Insp.By

1）准备好相关设备及软件。
1) Prepare equipment and software.
（1）Windows 7 及以上系统计算机一台；
（1）A computer with Windows 7 system or above;
（2）在计算机上安装有道 App 及 PDF 阅读器；
（2）Install Youdao App and PDF reader on the computer;
（3）在手机上安装有道 App。
（3）Install Youdao App on your mobile phone.

	工作者 Perf.By	检查者 Insp.By
2）选择有效的技术文件和手册。 2）Select effective technical documents and manuals.		
3）确认注意事项和警告的相关内容。 3）Confirm caution and warning.		

3. 工作步骤 Procedure	工作者 Perf.By	检查者 Insp.By
查询飞机 B-**** 空调区域灯不工作，自检维护信息为 "TEMP SELECTOR-LEFT PACK-NO GO" 故障相关跳开关、故障可能的原因等信息。 Query related trip switch, possible cause and other information when the air-conditioning lights of the aircraft B-**** are not Working and the self-check maintenance information is "TEMP SELECTOR-LEFT PACK-NO GO". * 请在手册中截出相应的图片内容说明每一步骤的查询结果 * *Please Use the Corresponding Screenshot in the Manual to Illustrate the Query Results of Each Step. * 查询步骤： Query Steps: 1. 查询飞机有效性（Effectivity） Check Aircraft Effectiveness 飞机 B-**** 的有效性代码（Effective Code）为 _____。 The effective code of B-**** is_____.		
2. 查询飞机空调区域灯不工作的系统章节（Chapter） Query The System Chapters Where the Lights in the Air-Conditioning Zone of the Aircraft are not Working 通过可观察故障清单（Observed Fault List）按系统（System-order）索引，飞机空调区域灯不工作的系统章节为 _____ 章。 Through the Observed Fault List according to the System-order index, the chapter of the system in which the lights in the air-conditioning of the aircraft does not work is _____.		
3. 确定飞机空调区域灯不工作，自检维护信息为 "TEMP SELECTOR-LEFT PACK-NO GO" 的故障隔离工卡 Confirm the FIM Task when the Lights in the Air-Conditioning of the Aircraft are not Working, and the BITE Maintenance Information is "TEMP SELECTOR-LEFT PACK-NO GO" 打开故障所在系统章节的自检维护信息列表，自检维护信息为 "TEMP SELECTOR-LEFT PACK-NO GO" 的故障隔离工卡号为 _____。 Open the BITE maintenance information list of the system chapter where the fault is located, and the FIM task number with the BITE maintenance information "TEMP SELECTOR-LEFT PACK-NO GO" is _____.		
4. 确定飞机空调区域灯不工作故障的相关跳开关信息 Determine the Relevant Jumper Information of the Aircraft Air-Conditioning Lights, not Working Failure 通过阅读飞机空调区域灯不工作故障隔离工卡可知故障相关跳开关信息。其中，相关跳开关 C00268 在面板 _____ 的 _____ 行、_____ 列。 By reading the fault isolation task of the aircraft air-conditioning lights not working, the fault-related circuit breaker information can be known. Circuit breaker C00268 is the _____ row and the _____ column of the panel _____.		

《民用航空飞机维修手册应用》实操工卡

班级 Class	工作卡号 Work Card No	FIM-05	共 3 页　第 3 页

5. 确定飞机空调区域灯不工作故障可能的原因 Determine the Possible Causes of the Failure about the Aircraft Air-Conditioning Lights 在飞机空调区域灯不工作故障隔离程序中，本故障最可能的原因是 ＿＿＿＿＿＿＿＿＿。 In the fault isolation procedure when the air-conditioning lights of the aircraft are not working, it can be seen that the most possible cause of this fault is ＿＿＿＿＿.	 飞机空调区域灯不工作 的故障隔离		
4．结束工作 **Close Out**		**工作者** **Perf.By**	**检查者** **Insp.By**
1）保存文件，文件用班级 + 学号 + 姓名 +FIM-05 的方式命名； 1）Save the file with the name of Class + Student ID + Name +FIM-05; 2）关机； 2）Shut down the computer; 3）6S 管理。 3）Conduct 6S Management.			

随手笔记

随手笔记

《民用航空飞机维修手册应用》实操工卡

工卡标题 Title	FIM 查询应用 6：飞机 PSEU 灯亮的故障隔离 FIM Query &Application 6: Fault Isolation in case of when the Aircraft PSEU Light being on			
机型 A/C Type	B737 6/7/8	难度等级 Difficulty Level		I
机号 REG.NO.	B-****	版本 Revision		R3
组别 Group		组长 Leader	组员 Team	学时 Period　1
参考资料 Reference	故障隔离手册 Fault Isolation Manual			
警告 Warning	请遵守将飞机置于空中模式的程序。在空中模式下，许多飞机系统都可以操作。这可能会造成人身伤害和设备损坏。 OBEY THE PROCEDURE THAT PREPARES TO PUT THE AIRPLANE IN THE AIR MODE. IN THE AIR MODE, MANY OF THE AIRPLANE SYSTEMS CAN OPERATE. THIS CAN CAUSE INJURIES TO PERSONS AND DAMAGE TO EQUIPMENT.			
编写人 Author		审核人 Reviewer		批准人 Approver
编写时间 Author Date		审核时间 Review Date		批准时间 Approve Date

工具 / 设备 / 材料 Tool/Equipment/Material				工作者 Perf.By	检查者 Insp.By
名称 Name	规格型号 Specification	数量 Quantity	使用情况 Usage		
计算机 Computer	Windows 7	1			
翻译软件 Translation App	有道等 Youdao etc.	1			
阅读器软件 Reader App	PDF 阅读器 PDF Reader App	1			

1. 工作任务 Requirement	工作者 Perf.By	检查者 Insp.By
飞机 PSEU 灯亮的故障隔离 Fault Isolation in case of when the Aircraft PSEU Light being on		

2. 工作准备 Job Set-up	工作者 Perf.By	检查者 Insp.By

1）准备好相关设备及软件。
1）Prepare equipment and software.
（1）Windows 7 及以上系统计算机一台；
（1）A computer with Windows 7 system or above;
（2）在计算机上安装有道 App 及 PDF 阅读器；
（2）Install Youdao App and PDF reader on the computer;
（3）在手机上安装有道 App。
（3）Install Youdao App on your mobile phone.

《民用航空飞机维修手册应用》实操工卡

	工作者 Perf.By	检查者 Insp.By
2）选择有效的技术文件和手册。 2）Select effective technical documents and manuals.		
3）确认注意事项和警告的相关内容。 3）Confirm caution and warning.		

3. 工作步骤 Procedure	工作者 Perf.By	检查者 Insp.By
查询飞机 B-****PSEU 灯亮，自检维护信息为 "AIR/GND R584 FLT" 的故障可能的原因、故障隔离程序等相关信息。 Query the relevant information such as possible cause when the B-****PSEU light is on, and the BITE maintenance information is "AIR/GND R584 FLT", and the fault isolation procedure, etc. * 请在手册中截出相应的图片内容说明每一步骤的查询结果 * *Please Use the Corresponding Screenshot in the Manual to Illustrate the Query Results of Each Step.* 查询步骤： Query Steps: 1. 查询飞机有效性（Effectivity） Check Aircraft Effectiveness 飞机 B-**** 的有效性代码（Effective Code）为 _____。 The effective code of B-**** is_____.		
2. 查询飞机 PSEU 灯亮的系统章节（Chapter） Query the System Chapter of the Aircraft PSEU Light on 通过可观察故障清单（Observed Fault List）按系统索引（System-order），飞机 PSEU 灯亮的系统章节为 _____ 章。 Through the Observed Fault List according to the System-order, the system chapter of the PSEU light of the aircraft is _____.		
3. 确定飞机 PSEU 灯亮，自检维护信息为 "AIR/GND R584 FLT" 的故障隔离工卡 Confirm the FIM Task when the Aircraft PSEU Light is on and the BITE Maintenance Information is "AIR/GND R584 FLT" 打开故障所在系统章节的自检维护信息列表，找到自检维护信息为 "AIR/GND R584 FLT" 的故障隔离工卡号为 _____。 Open the BITE maintenance information list of the system chapter where the fault is located, and the fault isolation task number with the BITE maintenance information of "AIR/GND R584 FLT" is _____.		
4. 确定飞机 PSEU 灯亮，自检维护信息为 "AIR/GND R584 FLT" 故障可能的原因 Confirm the Possible Cause of the fault when the Aircraft PSEU Light is on, and BITE Maintenance Information is "AIR/GND R584 FLT". 通过阅读本故障的故障隔离工卡，可知本故障最可能的原因是 _____。 By reading the fault isolation work card of this fault, we can know that the most possible cause of this fault is _____.		
5. 确定飞机 PSEU 灯亮，自检维护信息为 "AIR/GND R584 FLT" 的故障隔离程序 Confirm the Fault Isolation Procedure when the Aircraft PSEU Light is on and the BITE Maintenance Information is "AIR/GND R584 FLT" 通过阅读本故障的故障隔离工卡，可知故障隔离程序为 _____。 飞机 PSEU 灯亮的故障隔离 By reading the fault isolation work card of this fault, it can be known that the fault isolation procedure is _____.		

《民用航空飞机维修手册应用》实操工卡

4. 结束工作 Close Out	工作者 Perf.By	检查者 Insp.By
1）保存文件，文件用班级 + 学号 + 姓名 +FIM-06 的方式命名； 1）Save the file with the name of Class + Student ID + Name +FIM-06; 2）关机； 2）Shut down the computer; 3）6S 管理。 3）Conduct 6S Management.		

随手笔记

《民用航空飞机维修手册应用》实操工卡

工卡标题 Title	SSM 查询应用 1：飞机可收放着陆灯工作原理分析 SSM Query & Application 1: Working Principle Analysis of Aircraft Retractable Landing Lights		
机型 A/C Type	B737 6/7/8	难度等级 Difficulty Level	III
机号 REG.NO.	B-****	版本 Revision	R3

组别 Group		组长 Leader		组员 Team		学时 Period	1

参考资料 Reference	系统原理图手册 System Schematic Manual
警告 Warning	在进行组件更换之前，防止触申。断开跳开关可断开电源。在为照明系统提供电源，进行维护活动时发生意外接地可能会引起电击。 BEFORE YOU DO COMPONENT REPLACEMENTS,PROTECT YOURSELF FROM ELECTRICAL SHOCK. OPEN CIRCUIT BREAKERS TO REMOVE ELECTRICAL POWER. ACCIDENTAL GROUNDS DURING MAINTENANCE ACTIVITIES CAN CAUSE ELECTRICAL SHOCK WHEN ELECTRICAL POWER IS SUPPLIED TO THE LIGHTING SYSTEM.

编写人 Author		审核人 Reviewer		批准人 Approver	
编写时间 Author Date		审核时间 Review Date		批准时间 Approve Date	

工具 / 设备 / 材料 Tool/Equipment/Material				工作者 Perf.By	检查者 Insp.By
名称 Name	规格型号 Specification	数量 Quantity	使用情况 Usage		
计算机 Computer	Windows 7	1			
翻译软件 Translation App	有道等 Youdao etc.	1			
阅读器软件 Reader App	PDF 阅读器 PDF Reader App	1			

1. 工作任务 Requirement	工作者 Perf.By	检查者 Insp.By
飞机可收放着陆灯工作原理分析 Working Principle Analysis of Aircraft Retractable Landing Lights		

2. 工作准备 Job Set-up	工作者 Perf.By	检查者 Insp.By

1）准备好相关设备及软件。
1）Prepare equipment and software.
（1）Windows 7 及以上系统计算机一台；
（1）A computer with Windows 7 system or above;

（2）在计算机上安装有道 App 及 PDF 阅读器； （2）Install Youdao App and PDF reader on the computer; （3）在手机上安装有道 App。 （3）Install Youdao App on your mobile phone.		
2）选择有效的技术文件和手册。 2）Select effective technical documents and manuals.		
3）确认注意事项和警告的相关内容。 3）Confirm caution and warning.		
3. 工作步骤 **Procedure**	**工作者** **Perf.By**	**检查者** **Insp.By**
查询飞机 B-**** 可收放着陆灯原理图，分析飞机起飞和着陆时，开关打在"ON"位，"可收放着陆灯不放下，灯也不亮"的故障原因。 Query the schematic diagram of retractable landing light of the aircraft B-****, and analyze the cause of the failure when the switch is set to the "ON" position, "the retractable landing light will not be lowered and the light will not turn on" with the aircraft taking off and landing. * 请在手册中截出相应的图片内容说明每一步骤的查询结果 * *Please Use the Corresponding Screenshot in the Manual to Illustrate the Query Results of Each Step. * 查询步骤： Query Steps: 1. 查询飞机有效性（图略） Check Aircraft Effectiveness		
2. 查询 SDS 33 章灯光系统——着陆灯（Landing Lights）中可收放着陆灯的基本功能。 Query the system Function of the retractable landing lights in the chapter 33 of the SDS. **BOEING** 737-600/700/800/900 AIRCRAFT MAINTENANCE MANUAL LIGHTS - LANDING LIGHTS - RETRACTABLE LANDING LIGHTS **Purpose** The landing lights help the pilots see the runway during takeoff and landing. **Physical Description** The retractable landing light has these parts: • Lens assembly • Lamp • Retainer screws • Extend/retract motor. **Location** The retractable landing lights are on the fuselage, adjacent to the ram air inlet panels. The control switches for the retractable landing lights are on the P5 forward overhead panel. **Training Information Point** You must extend the retractable landing light to replace the lamp. After you extend the retractable landing light, pull and collar the circuit breaker to prevent accidental operation of retraction mechanism or electric shock.		

班级 Class	工作卡号 Work Card No	SSM-01	共 3 页　第 3 页	

3. 查询 SSM33-42-11 着陆灯（包括可收放着陆灯和固定着陆灯）工作原理图
Query the System Schematic Diagram of SSM33-42-11 Landing Lights（Including Retractable Landing Lights and Fixed Landing Lights）

4. 分析可收放着陆灯工作原理
Analyze the Working Principle of Retractable Landing Lights

通过分析可知，右侧可收放着陆灯由 1 号 115 交流汇流条供电，跳开关设备号为 C274，额定电流为 7.5 安。当右侧可收放着陆灯开关打在"ON"位时，S261 的 2 号与 3 号接通，如果可收放着陆灯不放下，灯也不亮，有五种可能性。详情见正文项目 4。

Through analysis, it can be seen that the retractable landing light on the right is powered by No. 1 115 AC bus, the number of the circuit breaker is C274, and the rated current is 7.5 amperes. When the retractable landing light switch on the right is in the "ON" position, No. 2 and No. 3 of S261 are connected. If the retractable landing light is not put down, the light does not turn on. There are five possibilities. See item 4 of the text for more details.

飞机可收放着陆灯
工作原理分析

4. 结束工作 Close Out	工作者 Perf.By	检查者 Insp.By
1）保存文件，文件用班级 + 学号 + 姓名 +SSM-01 的方式命名； 1）Save the file with the name of Class + Student ID + Name +SSM-01; 2）关机； 2）Shut down the computer; 3）6S 管理。 3）Conduct 6S Management.		

《民用航空飞机维修手册应用》实操工卡

班级 Class	工作卡号 Work Card No	SSM-02	共 2 页　第 1 页

工卡标题 Title	SSM 查询应用 2：飞机客舱空气循环系统的原理分析 SSM Query & Application 2: Principle Analysis of Air Circulation System in the Aircraft Cabin		
机型 A/C Type	B737 6/7/8	难度等级 Difficulty Level	II
机号 REG.NO.	B-****	版本 Revision	R3

组别 Group		组长 Leader		组员 Team		学时 Period	1

参考资料 Reference	系统原理图手册 System Schematic Manual
注意事项 Note	将手放在风扇上去感受振动。如果您没有感受到振动，则说明风扇已关闭。 PUT YOUR HAND ON THE FAN TO FEEL FOR VIBRATION. IF YOU DO NOT FEEL VIBRATION, THE FAN IS OFF.

编写人 Author		审核人 Reviewer		批准人 Approver	
编写时间 Author Date		审核时间 Review Date		批准时间 Approve Date	

工具 / 设备 / 材料 Tool/Equipment/Material				工作者 Perf.By	检查者 Insp.By
名称 Name	规格型号 Specification	数量 Quantity	使用情况 Usage		
计算机 Computer	Windows 7	1			
翻译软件 Translation App	有道等 Youdao etc.	1			
阅读器软件 Reader App	PDF 阅读器 PDF Reader App	1			

1. 工作任务 Requirement	工作者 Perf.By	检查者 Insp.By
飞机客舱空气循环系统的原理分析 Principle Analysis of Air Circulation System in the Aircraft Cabin		

2. 工作准备 Job Set-up	工作者 Perf.By	检查者 Insp.By

1）准备好相关设备及软件。
1）Prepare equipment and software.
（1）Windows 7 及以上系统计算机一台；
（1）A computer with Windows 7 system or above;
（2）在计算机上安装有道 App 及 PDF 阅读器；
（2）Install Youdao App and PDF reader on the computer;
（3）在手机上安装有道 App。
（3）Install Youdao App on your mobile phone.

2）选择有效的技术文件和手册。
2）Select effective technical documents and manuals.

《民用航空飞机维修手册应用》实操工卡

3）确认注意事项和警告的相关内容。 3）Confirm caution and warning.		
3. 工作步骤 **Procedure**	**工作者** **Perf.By**	**检查者** **Insp.By**
查询飞机 B-**** 客舱空气循环系统简图，分析该系统的组成 Query the Schematic Diagram of the Aircraft B-**** Cabin Air Circulation System and Analyze the Composition of the System * 请在手册中截出相应的图片内容说明每一步骤的查询结果 * *Please Use the Corresponding Screenshot in the Manual to Illustrate the Query Results of Each Step. * 查询步骤： Query Steps： 1. 查询飞机有效性（Effectivity） Check Aircraft Effectiveness 飞机 B-**** 的有效性代码（Effective Code）为 _____。 The effective code of B-**** is_____.		
2. 查询飞机客舱空气循环系统（Cabin Air Recirculation System）的功能 Inquire about the Function of Cabin Air Recirculation System AMM 的 SDS 空调系统（Air Conditioning）位于 _____ 章（Chapter），再循环系统（Recirculation System）位于 _____ 节（Section），客舱空气循环系统位于 _____ 目（Subject），飞机客舱空气循环系统的功能为 _____。 The chapter of Air Conditioning in SDS is _____; The section of Recirculation System is_____; The subject of Cabin Air Circulation is_____; the function of Aircraft Cabin Air Circulation System is_____.		
3. 查询飞机客舱空气循环系统简图 Query the Schematic Diagram of the Aircraft Cabin Air Circulation System 打开飞机客舱空气循环系统，可查得该系统简图。 Open the aircraft cabin air circulation system, you can find the system schematic diagram.		
4. 确定飞机客舱空气循环系统组成 Determine the Composition of the Aircraft Cabin Air Circulation System 根据飞机客舱空气循环系统简图，可知该系统组成包括 _____。 According to the schematic diagram of the aircraft cabin air circulation system, it can be seen that the system consists of_____.　　飞机客舱空气循环系统 的原理分析		
4. 结束工作 **Close Out**	**工作者** **Perf.By**	**检查者** **Insp.By**
1）保存文件，文件用班级＋学号＋姓名＋SSM-02 的方式命名； 1）Save the file with the name of Class + Student ID + Name +SSM-02; 2）关机； 2）Shut down the computer; 3）6S 管理。 3）Conduct 6S Management.		

《民用航空飞机维修手册应用》实操工卡

工卡标题 Title	SSM 查询应用 3：飞机空中交通管制（ATC）应答机的原理分析 SSM Query & Application 3: Principle Analysis of ATC Transponder			
机型 A/C Type	B737 6/7/8	难度等级 Difficulty Level		II
机号 REG.NO.	B-****	版本 Revision		R3
组别 Group		组长 Leader	组员 Team	学时 Period	1
参考资料 Reference	系统原理图手册 System Schematic Manual			
注意事项 Caution	请勿触摸 ATC 应答机上的连接器插针或其他导体。如果您触摸这些导体，则静电放电可能会对 ATC 转发器造成损坏。 DO NOT TOUCH THE CONNECTOR PINS OR OTHER CONDUCTORS ON THE ATCTRANSPONDERS. IF YOU TOUCH THESE CONDUCTORS, ELECTROSTATIC DISCHARGE CAN CAUSE DAMAGE TO THE ATC TRANSPONDERS.			
编写人 Author		审核人 Reviewer		批准人 Approver	
编写时间 Author Date		审核时间 Review Date		批准时间 Approve Date	

工具 / 设备 / 材料 Tool/Equipment/Material				工作者 Perf.By	检查者 Insp.By
名称 Name	规格型号 Specification	数量 Quantity	使用情况 Usage		
计算机 Computer	Windows 7	1			
翻译软件 Translation App	有道等 Youdao etc.	1			
阅读器软件 Reader App	PDF 阅读器 PDF Reader App	1			

1. 工作任务 Requirement	工作者 Perf.By	检查者 Insp.By
飞机空中交通管制（ATC）应答机的原理分析 Principle Analysis of ATC Transponder		

2. 工作准备 Job Set-up	工作者 Perf.By	检查者 Insp.By
1）准备好相关设备及软件。 1）Prepare equipment and software. （1）Windows 7 及以上系统计算机一台； （1）A computer with Windows 7 system or above; （2）在计算机上安装有道 App 及 PDF 阅读器； （2）Install Youdao App and PDF reader on the computer; （3）在手机上安装有道 App。 （3）Install Youdao App on your mobile phone.		

《民用航空飞机维修手册应用》实操工卡

班级 Class	工作卡号 Work Card No	SSM-03	共 2 页　第 2 页

2）选择有效的技术文件和手册。 2）Select effective technical documents and manuals.		
3）确认注意事项和警告的相关内容。 3）Confirm caution and warning.		

3. 工作步骤 Procedure	工作者 Perf.By	检查者 Insp.By
查询飞机 B-**** 空中交通管制（ATC）应答机的简图，分析该机载设备的组成 Query the Schematic Diagram of Air Traffic Control (Atc) Transponder, and Analyze the Composition of the Airborne Equipment * 请在手册中截出相应的图片内容说明每一步骤的查询结果 * *Please Use the Corresponding Screenshot in the Manual to Illustrate the Query Results of Each Step. * 查询步骤： Query Steps: 1. 查询飞机有效性（Effectivity） Check Aircraft Effectiveness 飞机 B-**** 的有效性代码（Effective Code）为＿＿＿＿＿＿。 The effective code of B-**** is＿＿＿＿＿＿.		
2. 查询飞机 ATC 应答机简图 (ATC Transponder) 的功能 Query the Function of the Aircraft ATC Transponder Diagram AMM 的 SDS 导航系统（Navigation）位于＿＿＿＿＿章（Chapter），空中交通管制系统（ATC System）位于＿＿＿＿＿节（Section），飞机 ATC 应答机＿＿＿＿＿目（Subject），飞机空中交通管制（ATC）应答机的功能为＿＿＿＿＿。 The chapter of Navigation in SDS is＿＿＿＿; The section of ATC System is＿＿＿＿; The subject of ATC Transponder is＿＿＿＿; the function of Aircraft ATC Transponder is＿＿＿＿.		
3. 查询飞机空中交通管制（ATC）应答机简图（1 号应答机和 2 号应答机任选）。 Query Aircraft Air Traffic Control (ATC) Transponder Diagram (Optional for No. 1 Transponder and No. 2 Transponder) 打开飞机空中交通管制（ATC）应答机，可查得该系统简图。 Turn on the aircraft air traffic control (ATC) transponder, and you can check the system schematic diagram.		
4. 确定飞机空中交通管制（ATC）应答机组成。 Determine the Composition of Aircraft Air Traffic Control (ATC) Transponder 根据飞机空中交通管制（ATC）应答机简图，可知 ATC 控制面板的图示为＿＿＿＿＿＿。 According to the schematic diagram of the aircraft air traffic control (ATC) transponder, it can be seen that the icon of the ATC control panel is＿＿＿＿＿. 飞机空中交通管制（ATC）应答机的原理分析		
4. 结束工作 Close Out	工作者 Perf.By	检查者 Insp.By
1）保存文件，文件用班级＋学号＋姓名＋SSM-03 的方式命名； 1）Save the file with the name of Class + Student ID + Name +SSM-03; 2）关机； 2）Shut down the computer; 3）6S 管理。 3）Conduct 6S Management.		

《民用航空飞机维修手册应用》实操工卡

工卡标题 Title	SSM 查询应用 4：飞机气象雷达冷却风扇的工作原理 SSM Query &Application 4: Working Principle of Airplane Weather Radar Fan			
机型 A/C Type	B737 6/7/8	难度等级 Difficulty Level		I
机号 REG.NO.	B-****	版本 Revision		R3
组别 Group		组长 Leader		组员 Team	学时 Period	1
参考资料 Reference	系统原理图手册 System Schematic Manual			
警告 Warning	当系统运行且天线罩打开时，请保持天气雷达天线的清洁。移动时，天线可以命中或抓住人员。天线传输微波能量时，请勿触摸天线表面。这些情况可能会导致人员受伤。 STAY CLEAR OF THE WEATHER RADAR ANTENNA WHEN THE SYSTEM IS IN OPERATION AND THE RADOME IS OPEN.THE ANTENNA CAN HIT OR CATCH PERSONNEL WHEN IT MOVES.DO NOT TOUCH THE ANTENNA SURFACE WHILE THE ANTENNA TRANSMITS MICROWAVE ENERGY.THESE CONDITIONS CAN CAUSE INJURIES TO PERSONNEL.			
编写人 Author		审核人 Reviewer		批准人 Approver		
编写时间 Author Date		审核时间 Review Date		批准时间 Approve Date		

工具 / 设备 / 材料 Tool/Equipment/Material				工作者 Perf.By	检查者 Insp.By
名称 Name	规格型号 Specification	数量 Quantity	使用情况 Usage		
计算机 Computer	Windows 7	1			
翻译软件 Translation App	有道等 Youdao etc.	1			
阅读器软件 Reader App	PDF 阅读器 PDF Reader App	1			
1. 工作任务 Requirement				工作者 Perf.By	检查者 Insp.By
飞机气象雷达冷却风扇的工作原理 Working Principle of Airplane Weather Radar Fan					
2. 工作准备 Job Set-up				工作者 Perf.By	检查者 Insp.By

1）准备好相关设备及软件。

1) Prepare equipment and software.

（1）Windows 7 及以上系统计算机一台；

(1) A computer with Windows 7 system or above;

（2）在计算机上安装有道 App 及 PDF 阅读器； （2）Install Youdao App and PDF reader on the computer; （3）在手机上安装有道 App。 （3）Install Youdao App on your mobile phone.	
2）选择有效的技术文件和手册。 2）Select effective technical documents and manuals.	
3）确认注意事项和警告的相关内容。 3）Confirm caution and warnings.	

3. 工作步骤 Procedure	工作者 Perf.By	检查者 Insp.By
查询飞机 B-**** 气象雷达冷却风扇的简图，分析该设备的工作原理 Query the Schematic Diagram of the Cooling Fan of the Aircraft B-**** Weather Radar and Analyze the Working Principle of the Equipment * 请在手册中截出相应的图片内容说明每一步骤的查询结果 * *Please Use the Corresponding Screenshot in the Manual to Illustrate the Query Results of Each Step. * 查询步骤： Query Steps: 1. 查询飞机有效性（Effectivity） Check Aircraft Effectiveness 飞机 B-**** 的有效性代码（Effective Code）为 _____。 The effective code of B-**** is_____.		
2. 查询飞机气象雷达冷却风扇的功能 Inquire about the Function of Weather Radar Cooling Fan AMM 的 SDS 导航系统（Navigation）位于 _____ 章（Chapter），气象雷达系统（Weather Radar System）位于 _____ 节（Section），气象雷达位于 _____ 目（Subject），飞机气象雷达冷却风扇的功能为 _____。 The chapter of Navigation in SDS is_____. The section of Weather Radar System is_____. The subject of Weather Radar is_____. the function of Weather Radar Cooling Fan is_____.		
3. 查询飞机气象雷达冷却风扇相关原理图 Query the Relevant Schematic Diagram of the Cooling Fan of the Aircraft Weather Radar 打开飞机气象雷达系统，可查得该系统简图。 Open the aircraft weather radar system, you can find the system diagram.		
4. 确定飞机气象雷达冷却风扇的工作原理 Determine the Working Principle of the Cooling Fan of the Aircraft Weather Radar 根据飞机气象雷达系统简图，分析飞机气象雷达冷却风扇的工作原理，气象雷达冷却风扇工作电压为多少？有几种工作情况？ According to the schematic diagram of the aircraft weather radar system, analyze the working principle of the cooling fan of the aircraft weather radar. What is the working voltage of the cooling fan of the weather radar? How many working conditions are there? 飞机气象雷达冷却风扇的工作原理分析		

班级 Class	工作卡号 Work Card No	SSM-04	共 3 页　第 3 页

4．结束工作 Close Out	工作者 Perf.By	检查者 Insp.By
1）保存文件，文件用班级＋学号＋姓名＋SSM-04 的方式命名； 1）Save the file with the name of Class + Student ID + Name +SSM-04; 2）关机； 2）Shut down the computer; 3）6S 管理。 3）Conduct 6S Management.		

随手笔记

《民用航空飞机维修手册应用》实操工卡

工卡标题 Title	SSM 查询应用 5：飞机再循环风扇的工作原理分析 SSM Query and Application 5: Analysis of the Working Principle of Aircraft Recirculation Fan			
机型 A/C Type	B737 6/7/8	难度等级 Difficulty Level		I
机号 REG.NO.	B-****	版本 Revision		R3
组别 Group		组长 Leader	组员 Team	学时 Period	1
参考资料 Reference	系统原理图手册 System Schematic Manual			
注意事项 Note	将手放在风扇上去感受振动。如果感受到振动，则表明风扇已打开。 PUT YOUR HAND ON THE FAN TO FEEL FOR VIBRATION. IF YOU FEEL VIBRATION, THE FAN IS ON.			
编写人 Author		审核人 Reviewer		批准人 Approver	
编写时间 Author Date		审核时间 Review Date		批准时间 Approve Date	

工具 / 设备 / 材料 Tool/Equipment/Material				工作者 Perf.By	检查者 Insp.By
名称 Name	规格型号 Specification	数量 Quantity	使用情况 Usage		
计算机 Computer	Windows 7	1			
翻译软件 Translation App	有道等 Youdao etc.	1			
阅读器软件 Reader App	PDF 阅读器 PDF Reader App	1			

1. 工作任务 Requirement	工作者 Perf.By	检查者 Insp.By
飞机再循环风扇的工作原理分析 Analysis of the Working Principle of Aircraft Recirculation Fan		

2. 工作准备 Job Set-up	工作者 Perf.By	检查者 Insp.By
1）准备好相关设备及软件。 1）Prepare equipment and software. （1）Windows 7 及以上系统计算机一台； （1）A computer with Windows 7 system or above; （2）在计算机上安装有道 App 及 PDF 阅读器； （2）Install Youdao App and PDF reader on the computer; （3）在手机上安装有道 App。 （3）Install Youdao App on your mobile phone.		

	工作者 Perf.By	检查者 Insp.By
2）选择有效的技术文件和手册。 2) Select effective technical documents and manuals.		
3）确认注意事项和警告的相关内容。 3) Confirm caution and warning.		

3．工作步骤 Procedure	工作者 Perf.By	检查者 Insp.By
查询飞机 B-**** 再循环风扇相关系统的简图，分析该设备的工作原理 Query the Schematic Diagram of the Aircraft B-**** Recirculation Fan Related System, and Analyze the Working Principle of the Equipment * 请在手册中截出相应的图片内容说明每一步骤的查询结果 * *Please Use the Corresponding Screenshot in the Manual to Illustrate the Query Results of Each Step. * 查询步骤： Query Steps: 1．查询飞机有效性（Effectivity） Check Aircraft Effectiveness 飞机 B-**** 的有效性代码（Effective Code）为 _____。 The effective code of B-**** is_____.		
2．查询飞机再循环风扇（Recirculation Fan）的功能 Inquire about the Function of Recirculation Fan AMM 的 SDS 空调系统（Air Conditioning）位于 _____ 章（Chapter），再循环系统（Recirculation System）位于 _____ 节（Section），客舱空气循环系统（Cabin Air Recirculation System）位于 _____ 目（Subject），飞机再循环风扇的功能为 _____。 The chapter of Air conditioning in SDS is_____; The section of Recirculation System is_____; The subject of Cabin Air Recirculation System is_____; the function of Aircraft Recirculation Fan is_____.		
3．查询飞机再循环风扇相关原理图 Inquire about the Schematic Diagram of the Aircraft Recirculation Fan 打开再循环系统，可查得飞机再循环风扇相关原理图。 Open the recirculation system, you can find the relevant schematic diagram of the aircraft recirculation fan.		
4．确定飞机再循环风扇的工作原理 Determine the Working Principle of the Aircraft Recirculation Fan 根据再循环系统简图，分析飞机再循环风扇的工作原理，回答当再循环风扇电门 AUTO 位时，（1）左右组件电门 AUTO 位，该设备是否工作？（2）当一个组件关闭时，该设备是否能工作？ According to the schematic diagram of the recirculation system, analyze the working principle of the recirculation fan of the aircraft, and answer when the recirculation fan switch is AUTO position, (1) The left and right component switches are AUTO position, does the device work? (2) When a component is closed, can the device work? 飞机再循环风扇 工作原理分析		

《民用航空飞机维修手册应用》实操工卡

班级 Class	工作卡号 Work Card No	SSM-05	共 3 页 第 3 页

4. 结束工作 Close Out	工作者 Perf.By	检查者 Insp.By
1）保存文件，文件用班级 + 学号 + 姓名 +SSM-05 的方式命名； 1）Save the file with the name of Class + Student ID + Name +SSM-05; 2）关机； 2）Shut down the computer; 3）6S 管理。 3）Conduct 6S Management.		

随手笔记

随手笔记

《民用航空飞机维修手册应用》实操工卡

工卡标题 Title	SSM 查询应用 6：飞机厕所烟雾探测器的工作原理分析 SSM Query & Application 6: Working Principle Analysis of Airplane Lavatory Smoke Detectors		
机型 A/C Type	B737 6/7/8	难度等级 Difficulty Level	I
机号 REG.NO.	B-****	版本 Revision	R3

组别 Group		组长 Leader		组员 Team		学时 Period	1

参考资料 Reference	系统原理图手册 System Schematic Manual
警告 Warning	清洁烟雾探测器和传感器时，确保该区域有足够的气流。保持酒精远离热、火花或火焰。请勿从酒精中燃烧烟雾。戴上橡胶手套。酒精是一种易燃、有毒的溶剂，可能导致人身伤害和设备损坏。 WHEN YOU CLEAN THE SMOKE DETECTOR AND SENSOR, MAKE SURE THE AREA HAS SUFFICIENT AIRFLOW.KEEP THE ALCOHOL AWAY FROM HEAT, SPARKS OR FLAMES.DO NOT BREATH THE FUMES FROM THE ALCOHOL.[PUT ON RUBBER GLOVES]. ALCOHOL IS A FLAMMABLE AND POISONOUS SOLVENT WHICH CAN CAUSE INJURIES TO PERSONS AND DAMAGE TO EQUIPMENT.

编写人 Author		审核人 Reviewer		批准人 Approver	
编写时间 Author Date		审核时间 Review Date		批准时间 Approve Date	

工具 / 设备 / 材料 Tool/Equipment/Material				工作者 Perf.By	检查者 Insp.By
名称 Name	规格型号 Specification	数量 Quantity	使用情况 Usage		
计算机 Computer	Windows 7	1			
翻译软件 Translation App	有道等 Youdao etc.	1			
阅读器软件 Reader App	PDF 阅读器 PDF Reader App	1			

1. 工作任务 Requirement	工作者 Perf.By	检查者 Insp.By
飞机厕所烟雾探测器的工作原理分析 Working Principle Analysis of Airplane Lavatory Smoke Detectors		

2. 工作准备 Job Set-up	工作者 Perf.By	检查者 Insp.By
1）准备好相关设备及软件。 1）Prepare equipment and software.		

《民用航空飞机维修手册应用》实操工卡

	工作者	检查者
（1）Windows 7 及以上系统计算机一台； （1）A computer with Windows 7 system or above; （2）在计算机上安装有道 App 及 PDF 阅读器； （2）Install Youdao App and PDF reader on the computer; （3）在手机上安装有道 App。 （3）Install Youdao App on your mobile phone.		
2）选择有效的技术文件和手册。 2）Select effective technical documents and manuals.		
3）确认注意事项和警告的相关内容。 3）Confirm caution and warning.		

3. 工作步骤 Procedure	工作者 Perf.By	检查者 Insp.By
查询飞机 B-**** 前厕所烟雾探测器简图，分析该设备的工作原理 Query the Schematic Diagram of the Smoke Detector in the Front Lavatory of B-****, and Analyze the Working Principle of the Device * 请在手册中截出相应的图片内容说明每一步骤的查询结果 * *Please Use the Corresponding Screenshot in the Manual to Illustrate the Query Results of Each Step. * 查询步骤： Query Steps: 1. 查询飞机有效性（Effectivity） Check Aircraft Effectiveness 飞机 B-**** 的有效性代码（Effective Code）为 _____。 The effective code of B-**** is_____.		
2. 查询飞机厕所烟雾探测器的功能 Inquire about the Function of the Smoke Detector in the Lavatory AMM 的 SDS 防火系统（Fire Protection）位于 _____ 章（Chapter），厕所烟雾探测器（Lavatory Smoke Detectors）位于 _____ 节（Section），厕所烟雾探测器位于 _____ 目（Subject），厕所烟雾探测器的功能为 _____。 The chapter of Fire Protection in SDS is_____; The section of Lavatory Smoke Detector System is_____; The subject of Lavatory Smoke Detector is_____; the function of Lavatory Smoke Detector is_____.		
3. 查询厕所烟雾探测器相关原理图 Query the Related Schematic Diagrams of Toilet Smoke Detectors 打开飞机厕所烟雾探测器可查得该设备原理图。 Open the aircraft lavatory smoke detector to find the schematic diagram of the device.		
4. 确定厕所烟雾探测器的工作原理。 Determine the Working Principle of the Toilet Smoke Detector 根据厕所烟雾探测器原理图，分析厕所烟雾探测器的工作原理，则该设备工作电压为多少？有哪些组成部分？ According to the schematic diagram of the lavatory smoke detector analyze the working principle of the lavatory smoke detector, What is the working voltage of the device? What are the components? 飞机前厕所烟雾探测器的工作原理分析		

班级 Class	工作卡号 Work Card No	SSM-06	共 3 页　第 3 页

4. 结束工作 Close Out	工作者 Perf.By	检查者 Insp.By
1）保存文件，文件用班级 + 学号 + 姓名 +SSM-06 的方式命名； 1）Save the file with the name of Class + Student ID + Name +SSM-06; 2）关机； 2）Shut down the computer; 3）6S 管理。 3）Conduct 6S Management.		

随手笔记

随手笔记

《民用航空飞机维修手册应用》实操工卡

班级 Class		工作卡号 Work Card No	WDM-01		共 5 页　第 1 页	

工卡标题 Title	WDM 查询应用 1：襟翼位置传感器线路故障的排除 WDM Query & Application 1: Wiring Troubleshooting of Flap Position Sensor Wiring Fault			
机型 A/C Type	B737 6/7/8	难度等级 Difficulty Level		III
机号 REG.NO.	B-****	版本 Revision		R3

组别 Group		组长 Leader		组员 Team		学时 Period	

参考资料 Reference	线路图手册 Wiring Diagram Manual 故障隔离手册 Fault Isolation Manual
警告 Warning	保持人员和设备远离飞行控制表面。当您供应液压动力时，副翼、升降机、舵、襟翼、履带板、扰流板、稳定器和前齿轮会突然移动。这可能会造成人身伤害和设备损坏。 KEEP PERSONS AND EQUIPMENT AWAY FROM THE FLIGHT CONTROL SURFACES. THE AILERONS, ELEVATORS, RUDDER, FLAPS, SLATS, SPOILERS, STABILIZER AND NOSE GEAR CAN MOVE SUDDENLY WHEN YOU SUPPLY HYDRAULIC POWER. THIS CAN CAUSE INJURY TO PERSONS AND DAMAGE TO EQUIPMENT.

编写人 Author		审核人 Reviewer		批准人 Approver	
编写时间 Author Date		审核时间 Review Date		批准时间 Approve Date	

工具 / 设备 / 材料 Tool/Equipment/Material				工作者 Perf.By	检查者 Insp.By
名称 Name	规格型号 Specification	数量 Quantity	使用情况 Usage		
计算机 Computer	Windows 7	1			
翻译软件 Translation App	有道等 Youdao etc.	1			
阅读器软件 Reader App	PDF 阅读器 PDF Reader App	1			

1. 工作任务 Requirement	工作者 Perf.By	检查者 Insp.By
襟翼位置传感器线路故障的排除 Wiring Troubleshooting of Flap Position Sensor Wiring Fault		

2. 工作准备 Job Set-up	工作者 Perf.By	检查者 Insp.By

1）准备好相关设备及软件。
1) Prepare equipment and software.
（1）Windows 7 及以上系统计算机一台；
（1）A computer with Windows 7 system or above;
（2）在计算机上安装有道 App 及 PDF 阅读器；
（2）Install Youdao App and PDF reader on the computer;
（3）在手机上安装有道 App。
（3）Install Youdao App on your mobile phone.

2）选择有效的技术文件和手册。 2）Select effective technical documents and manuals.		
3）确认注意事项和警告的相关内容。 3）Confirm caution and warning.		

3. 工作步骤 Procedure	工作者 Perf.By	检查者 Insp.By
飞机 B-**** 襟翼位置传感器线路故障导致自动飞行控制系统 B 中 "SNSR EXC AC" 跳开关跳出，学习并模仿该线路故障的排除。 Aircraft B-**** flap position sensor circuit failure caused the "SNSR EXC AC" switch in automatic flight control system B to jump out, learn to imitate the troubleshooting of the circuit failure. * 请在手册中截出相应的图片内容说明每一步骤的查询结果 * *Please Use the Corresponding Screenshot in the Manual to Illustrate the Query Results of Each Step. * 查询步骤： Query Steps: 1. 查询飞机的有效性（图略） Check the Validity of the Aircraft (Picture Omitted)		
2. 查询故障隔离手册（FIM）22 章自动飞行控制系统可获得该故障可能的原因及故障隔离程序 Consult Auto Flight Control System in Chapter 22 of the Fault Isolation Manual (FIM) to Obtain the Possible Causes of the Fault and the Fault Isolation Procedures		

BOEING
737-600/700/800/900
FAULT ISOLATION MANUAL

807. **Flap Position Transmitter Problem - Fault Isolation**

A. Description

  (1) This task is for these maintenance messages:

    NOTE: J1A-XX, J1B-XX and J1C-XX (where XX is a pin number) are the connector and pin numbers on the FCC-A and FCC-B. For the FCC-A: J1A = D10135A, J1B = D10135B and J1C = D10135C. For the FCC-B: J1A = 10137A, J1B = 10137B and J1C = 10137C.

    (a) FLAP POSN-1 (J1B-G06, F06, E08)

    (b) FLAP POSN-2 (J1B-G06, F06, E06)

  (2) FLAP POSN-1 (J1B-G06, F06, E06): The flight control computer A (FCC-A) receives incorrect data from the left flap position transmitter during the ground test.

  (3) FLAP POSN-2 (J1B-G06, F06, E06): The flight control computer B (FCC-B) receives incorrect data from the right flap position transmitter during the ground test.

B. Possible Causes

  (1) Left flap position transmitter, T427

  (2) Right flap position transmitter, T428

  (3) Flight control computer A (FCC-A), M1875

  (4) Flight control computer B (FCC-B), M1876

  (5) Wiring Problem

C. Circuit Breakers

  (1) These are the primary circuit breakers related to the fault:

    CAPT Electrical System Panel, P18-1

Row	Col	Number	Name
C	5	C01041	AFCS SYS A SNSR EXC AC

    F/O Electrical System Panel, P6-2

Row	Col	Number	Name
C	2	C01042	AFCS SYS B SNSR EXC AC

D. Related Data

  (1) (SSM 22-11-11)

  (2) (SSM 22-12-31)

  (3) (SSM 22-12-41)

  (4) (WDM 22-11-11)

  (5) (WDM 22-12-31)

  (6) (WDM 22-12-41)

  (4) If the maintenance message is on channel B, then do these steps:

    (a) Disconnect the right flap position transmitter connector, D229.

    NOTE: You can find the left flap position transmitter on the outboard end of the transmission No. 8.

    (b) Do a continuity check between the pins shown below, at the flap position transmitter, T428.

D229　　　　　　　　　　　　D229 　pin 9　------------------　pin 10 　pin 12　-----------------　pin 13 　pin 12　-----------------　pin 14 　pin 13　-----------------　pin 14 (c)　If there is an open circuit, do these steps: 　　1)　Replace the flap position transmitter, T428. 　　　　These are the tasks: 　　　　Flap Position Transmitter Removal, AMM TASK 27-58-01-000-801, 　　　　Flap Position Transmitter Installation, AMM TASK 27-58-01-400-801. 　　2)　Do the Repair Confirmation at the end of this task. (d)　If there is continuity between the above pins, then continue. (e)　Make sure that this circuit breaker is closed: 　　F/O Electrical System Panel, P6-2 　　Row　Col　Number　　Name 　　C　　2　　C01042　　AFCS SYS B SNSR EXC AC			
3. 查询 WDM 手册 22 章自动飞行控制系统（AUTO FLIGHT）的 22-11-11，查询 D229 的 Pin9、Pin10、Pin12、Pin13、Pin14 的相关线路 Query 22-11-11 of the Automatic Flight Control System in Chapter 22 of WDM, to Inquire about the Related Lines of Pin9, Pin10, Pin12, Pin13, and Pin14 Of D229 			
4. 分析襟翼位置传感器线路图 Analyze the Wiring Diagram of the Flap Position Sensor 			

班级 Class	工作卡号 Work Card No	WDM-01	共 5 页　第 4 页

5. 打开导线清单（Wire List），查询导线 W1034-2005B-20 可以得到连接设备的信息和导线类型代码

Open the Wire List and Query the Wire W1034-2005B-20 to Get the Information of the Connected Device and the Wire Type Code

◆BOEING   **737-700 WIRING DIAGRAM MANUAL**

Bundle No. Wire No.	GA	CO TY	Fam	Description FT-IN	Diagram	From Equip	Term	Type	Splice	To Equip	Term	Type	Splice	Effectivity
W1032		286A1032			R WING TRAILNG EDGE:DISC BRACKET STA 254R TO W/W DISC & AFT CARGO DISC (continued)									
3011R	20	Y7	WK	0-0	27-62-14	D40034P	44			D43100P	16			YA811-YA814
3011Y	20	Y7	WK	0-0	27-62-14	D40034P	46			D43100P	18			YA811-YA814
3011Z			WK	0-0	27-62-14	D40034P	S/R			D43100P	S/R			YA811-YA814
W1034		286A1034			RIGHT WING TRAILING EDGE:STA 525 TO R. WING DISC STA 254R									
0513	20	HL		1-5	27-52-11	D00229	15			GD00136	ST..	E		ALL
1001	20	HP	AA	16-11	28-21-11	D00790	6			D40034J	34			ALL
1001Z			AA	0-0	28-21-11	D00790	S/R			D40034J	S/R			ALL
1002	18	HP	AB	2-8	27-53-21	D11826	8			GD00136	ST..	E	*1	ALL
1002Z			AB	0-0	27-53-21	D11826	S/R			GD00136	ST..	E	*1	ALL
1003	18	HP	AC	8-11	27-53-21	D11826	8			GD00134	ST..	E	*2	ALL
1003Z			AC	0-0	27-53-21	D11826	S/R			GD00134	ST..	E	*2	ALL
2001B	20	Y6	GA	28-0	27-18-11	D03574	1			D40034J	7			ALL
2001R	20	Y6	GA	0-0	27-18-11	D03574	2			D40034J	6			ALL
2001Z			GA	0-0	27-18-11	G000130	ST..	E		D40034J	S/R			ALL
2002B	20	Y6	GB	13-1	22-11-31	D01697	2			D40034J	2			ALL
2002R	20	Y6	GB	0-0	22-11-31	D01697	1			D40034J	1			ALL
2002Z			GB	0-0	22-11-31	D01697	S/R			D40034J	S/R			ALL
2003B	20	Y6	GC	13-1	22-11-31	D01701	2			D40034J	20			ALL
2003R	20	Y6	GC	0-0	22-11-31	D01701	1			D40034J	19			ALL
2003Z			GC	0-0	22-11-31	D01701	S/R			D40034J	S/R			ALL
2004B	20	Y6	GD	20-10	27-32-11	D00229	5			D40034J	30			ALL
2004R	20	Y6	GD	0-0	27-32-11	D00229	4			D40034J	29			ALL
2004Z			GD	0-0	27-32-11	D00229	S/R			D40034J	S/R			ALL
2005B	20	Y6	GE	20-10	22-11-11	D00229	10			D40034J	25			ALL
2005R	20	Y6	GE	0-0	22-11-11	D00229	9			D40034J	24			ALL
2005Z			GE	0-0	22-11-11	D00229	S/R			D40034J	S/R			ALL
2006B	20	M2	GF	20-10	27-52-11	D00229	2			D40034J	37			ALL
2006R	20	M2	GF	0-0	27-52-11	D00229	1			D40034J	36			ALL
2006ZI			GF	0-0	27-52-11	D00229	S/R			D40034J	38			ALL
2006ZO			GF	0-0	27-52-11	D00229	S/R			D40034J	S/R			ALL
2007B	20	M2	GG	20-10	27-52-11	D00229	3			D40034J	41			ALL
2007R	20	M2	GG	0-0	27-52-11	D00229	11			D40034J	40			ALL
2007ZI			GG	0-0	27-52-11	D00229	S/R			D40034J	39			ALL
2007ZO			GG	0-0	27-52-11	D00229	S/R			D40034J	S/R			ALL
2008B	20	Y6	GI	18-5	27-53-21	D11826	7			D11826	3			ALL
2008R	20	Y6	GI	0-0	27-53-21	D11826	6			D11826	2			ALL
2008Z			GI	0-0	27-53-21	D11826	S/R			D11826	S/R			ALL
2009B	20	Y6	GH	11-10	27-53-21	D11826	1			D40034J	12			ALL

6. 打开 WDM 的设备清单（Equipment List），查询设备件号等信息

Open the WDM Equipment List, and Query the Equipment Part Number and Other Information

◆BOEING   **737-700 WIRING DIAGRAM MANUAL**

Equip	Opt	Part Number Part Description	Used On Dwg Vendor	Qty	Diagram Station / WL / BL	Effectivity
D00201	1	BACC66F11F04AE01 PLUG-VHF-2 TRANSCEIVER	81205	1	23-12-21 E001-05/ /	ALL
D00201 T	1	S280W601-116 GROUND BLOCK, ARINC 600	81205	1	23-12-21 "/"/	ALL
D00201A	1	PARTAOFD00201 PLUG-VHF-2 TRANSCEIVER		0	23-12-21 "/"/	ALL
D00201B	1	PARTBOFD00201 PLUG-VHF-2 TRANSCEIVER		0	23-12-21 "/"/	ALL
D00201C	1	PARTCOFD00201 PLUG-VHF-2 TRANSCEIVER		0	23-12-21 "/"/	ALL
D00203	1	BACC63CB16-24SN PLUG-CONDUCTIVE CONNECTOR	81205	1	23-12-11 P8-2/ /	YA811-YA814
D00203 T	1	BACC10KD16 BACKSHELL-CONDUCTIVE	81205	1	23-12-11 P8-2/ /	YA811-YA814
D00209	1	BACC63CB16-24SN PLUG-CONDUCTIVE CONNECTOR	81205	1	23-12-21 P8-3/ /	ALL
D00209 T	1	BACC10KD16 BACKSHELL-CONDUCTIVE	81205	1	23-12-11 P8-3/ /	ALL
D00211	1	BACC45FT16-24S7 PLUG-ATC DUAL CONT	81205	1	34-53-11 P008-29/ /	ALL
D00211 T	1	BACC10JS16 BACKSHELL	81205	1	34-53-11 "/"/	ALL
D00229	1	BACC63BP20C16SN PLUG-CONDUCTIVE	81205	1	27-52-11 ORSS/446/	ALL

7. 查询设备所有连线信息

Query All Connection Information of the Device

打开 WDM 的连线清单（Hook-Up List），查询 D229 的连线信息。

Open the Hook-Up List of WDM and query the connection information of D229.

襟翼位置传感器线路
故障的排除思路

班级 Class	工作卡号 Work Card No	WDM-01	共 5 页　第 5 页

737-700 WIRING DIAGRAM MANUAL

Equip	Station / WL / BL Term Type	Bundle	Description Wire No.	GA	CO	Diagram	Effectivity
D00211	P008-29/ /		PLUG-ATC DUAL CONT (continued)				ALL
	5　VP	W2177	0005	20		34-53-11	ALL
	6	W2177	0006	20		34-53-11	ALL
	7	W2177	0007	24		34-53-11	ALL
	8	W2177	0008	20		34-53-11	ALL
	9　VP	W2177	0009	20		34-53-11	ALL
	10　VP	W2177	0010	20		34-53-11	ALL
	11	W2177	0011	24		34-53-11	ALL
	12	W2177	0012	24		34-53-11	ALL
	13　UNUSED						ALL
	14　UNUSED						ALL
	15	W2177	0015	24		34-53-11	ALL
	16	W2177	0016	24		34-53-11	ALL
	17　UNUSED						ALL
	18	W2177	0018	20		33-18-61	ALL
	19　UNUSED						ALL
	20	W2177	0020	24		34-53-11	ALL
	21	W2177	0021	20		33-18-61	ALL
	22	W2177	2001R	24		34-53-11	ALL
	23	W2177	2001B	24		34-53-11	ALL
	24	W2177	0024	24		34-53-11	ALL
	DED	W2177	2001Z			34-53-11	ALL
D00229	ORSS/446/		PLUG-CONDUCTIVE				ALL
	1	W1034	2006R	20		27-52-11	ALL
	2	W1034	2006B	20		27-52-11	ALL
	3	W1034	2007B	20		27-52-11	ALL
	4	W1034	2004R	20		27-32-11	ALL
	5	W1034	3004R	20		27-32-11	ALL
	6	W1034	3004R	20		27-32-11	ALL
	7	W1034	3004B	20		27-32-11	ALL
	8	W1034	3004Y	20		27-32-11	ALL
	9	W1034	2005R	20		22-11-11	ALL
	10	W1034	2005B	20		22-11-11	ALL
	11	W1034	2007R	20		27-52-11	ALL
	12	W1034	3005R	20		22-12-41	ALL
	13	W1034	3005B	20		22-12-41	ALL
	14	W1034	3005Y	20		22-12-41	ALL
	15	W1034	0513	20		27-52-11	ALL
	16　UNUSED						ALL
	S/R	W1034	2004Z			27-32-11	ALL

4. 结束工作 Close Out	工作者 Perf.By	检查者 Insp.By
1）保存文件，文件用班级 + 学号 + 姓名 +WDM-01 的方式命名； 1）Save the file with the name of Class + Student Id + Name + WDM-01; 2）关机； 2）Shut down the computer; 3）6S 管理。 3）Conduct 6S Management.		

随手笔记

《民用航空飞机维修手册应用》实操工卡

班级 Class		工作卡号 Work Card No	WDM-02	共 2 页　第 1 页	

工卡标题 Title	WDM 查询应用 2：飞机上设备 M1020 的线路分析 WDM Query & Application 2: Wiring Analysis of the Equipment M1020 in the Aircraft				
机型 A/C Type	B737 6/7/8		难度等级 Difficulty Level		II
机号 REG.NO.	B-****		版本 Revision		R3
组别 Group		组长 Leader		组员 Team	学时 Period
参考资料 Reference	线路图手册 Wiring Diagram Manual 零部件图解目录手册 Illustrated Parts Catalog 飞机维护手册 Aircraft Maintenance Manual				
注意事项 Note	A/P 副翼执行器 A 和 A/P 升降舵执行器 A 与飞行控制计算机 A（通道 A）有接口。A/P 副翼执行器 B 和 A/P 升降舵执行器 B 与飞行控制计算机 B（通道 B）有接口。 A/P AILERON ACTUATOR A AND A/P ELEVATOR ACTUATOR A HAVE INTERFACES WITH THE FLIGHT CONTROL COMPUTER A (CHANNEL A). A/P AILERON ACTUATOR B AND A/P ELEVATOR ACTUATOR B HAVE INTERFACES WITH THE FLIGHT CONTROL COMPUTER B (CHANNEL B).				
编写人 Author		审核人 Reviewer		批准人 Approver	
编写时间 Author Date		审核时间 Review Date		批准时间 Approve Date	

工具 / 设备 / 材料 Tool/Equipment/Material				工作者 Perf.By	检查者 Insp.By
名称 Name	规格型号 Specification	数量 Quantity	使用情况 Usage		
计算机 Computer	Windows 7	1			
翻译软件 Translation App	有道等 Youdao etc.	1			
阅读器软件 Reader App	PDF 阅读器 PDF Reader App	1			

1. 工作任务 Requirement	工作者 Perf.By	检查者 Insp.By
飞机上设备 M1020 的线路分析 Wiring Analysis of the Equipment M1020 in the Aircraft		

2. 工作准备 Job Set-up	工作者 Perf.By	检查者 Insp.By
1）准备好相关设备及软件。 1）Prepare equipment and software. （1）Windows 7 及以上系统计算机一台； （1）A computer with Windows 7 system or above; （2）在计算机上安装有道 App 及 PDF 阅读器； （2）Install Youdao App and PDF reader on the computer; （3）在手机上安装有道 App。 （3）Install Youdao App on your mobile phone.		

班级 Class	工作卡号 Work Card No	WDM-02	共 2 页　第 2 页

2）选择有效的技术文件和手册。 2）Select effective technical documents and manuals.		
3）确认注意事项和警告的相关内容。 3）Confirm caution and warning.		

3. 工作步骤 **Procedure**	工作者 Perf.By	检查者 Insp.By
查询飞机 B-**** 上设备 M1020 的件号、线路连接等信息 Search for the Part Number, Line Connection and Other Information of the Equipment M1020 on the Aircraft B-**** * 请在手册中截出相应的图片内容说明每一步骤的查询结果 * *Please Use the Corresponding Screenshot in the Manual to Illustrate the Query Results of Each Step. * 查询步骤： Query Steps: 1. 查询飞机有效性 Check Aircraft Effectiveness 飞机 B-**** 的有效性代码（Effective Code）为 _____。 The effective code of B-**** is_____.		
2. 查询飞机上设备 M1020 的件号 Query the Part Number of the Equipment M1020 on the Aircraft 查询 WDM 的设备清单（Equipment List），设备 M1020（M01020）的件号为 _____，设备所在的线路图章节号为 _____。 Search the WDM equipment list, the part number of the equipment M1020 (M01020) is _____, and the Chapter-Section-Subject of the wiring diagram of the equipment is _____.		
3. 确定飞机上设备 M1020 的线路连接信息 Confirm the Line Connection Information of the Equipment M1020 on the Plane 查询设备 M1020 所在的线路图纸，可确定该设备的线路连接信息。 Query the line drawing where the equipment M1020 is located to determine the line connection information of the equipment. 飞机上设备 M1020 的线路分析		

4. 结束工作 **Close Out**	工作者 Perf.By	检查者 Insp.By
1）保存文件，文件用班级＋学号＋姓名＋WDM-02 的方式命名； 1）Save the file with the name of Class + Student Id + Name + WDM-02; 2）关机； 2）Shut down the computer; 3）6S 管理。 3）Conduct 6S Management.		

《民用航空飞机维修手册应用》实操工卡

工卡标题 Title	WDM 查询应用 3：飞机上跳开关 C00413 的线路分析 WDM Query & Application 3: Wiring Analysis of the Circuit Breaker C00413 in the Aircraft		
机型 A/C Type	B737 6/7/8	难度等级 Difficulty Level	II
机号 REG.NO.	B-****	版本 Revision	R3

组别 Group		组长 Leader		组员 Team		学时 Period	1

参考资料 Reference	线路图手册 Wiring Diagram Manual 零部件图解目录手册 Illustrated Parts Catalog 飞机维护手册 Aircraft Maintenance Manual
注意事项 Note	跳开关列表可在 AMM 简介中找到。 THE LIST OF CIRCUIT BREAKERS CAN BE FOUND IN THE AMM INTRODUCTION.

编写人 Author		审核人 Reviewer		批准人 Approver	
编写时间 Author Date		审核时间 Review Date		批准时间 Approve Date	

工具 / 设备 / 材料 Tool/Equipment/Material				工作者 Perf.By	检查者 Insp.By
名称 Name	规格型号 Specification	数量 Quantity	使用情况 Usage		
计算机 Computer	Windows 7	1			
翻译软件 Translation App	有道等 Youdao etc.	1			
阅读器软件 Reader App	PDF 阅读器 PDF Reader App	1			

1. 工作任务 Requirement	工作者 Perf.By	检查者 Insp.By
飞机上跳开关 C00413 的线路分析 Wiring Analysis of the Circuit Breaker C00413 in the Aircraft		

2. 工作准备 Job Set-up	工作者 Perf.By	检查者 Insp.By
1）准备好相关设备及软件。 1）Prepare equipment and software. （1）Windows 7 及以上系统计算机一台； （1）A computer with Windows 7 system or above; （2）在计算机上安装有道 App 及 PDF 阅读器； （2）Install Youdao App and PDF reader on the computer; （3）在手机上安装有道 App。 （3）Install Youdao App on your mobile phone.		

班级 Class	工作卡号 Work Card No	WDM-03	共 2 页　第 2 页

	工作者 Perf.By	检查者 Insp.By
2）选择有效的技术文件和手册。 2）Select effective technical documents and manuals.		
3）确认注意事项和警告的相关内容。 3）Confirm caution and warning.		
3. 工作步骤 **Procedure**	工作者 Perf.By	检查者 Insp.By
查询飞机 B-**** 上跳开关 C00413 的件号、供电的电源等信息 Search for the Part Number, the Power Supply etc. of the Circuit Breaker C00413 on the Aircraft B-**** * 请在手册中截出相应的图片内容说明每一步骤的查询结果 * *Please Use the Corresponding Screenshot in the Manual to Illustrate the Query Results of Each Step. * 查询步骤： Query Steps: 1. 查询飞机有效性 Check Aircraft Effectiveness 飞机 B-**** 的有效性代码（Effective Code）为 _____。 The effective code of B-**** is_____.		
2. 查询飞机上跳开关 C00413 的件号 Query the Part Number of the Circuit Breaker C00413 on the Aircraft 通过 WDM 的设备清单（Equipment List），跳开关 C00413 的件号为 _____，设备所在的线路图章节号 _____。 Search the WDM equipment list, the part number of the circuit Breaker C00413 is _____, and the Chapter-Section-Subject of the wiring diagram of the equipment is _____.		
3. 确定飞机上跳开关 C00413 的供电电源信息 Confirm the Power Supply Information of the Circuit Breaker C00413 on the Aircraft 查询跳开关 C00413 所在的线路图纸，可确定跳开关的供电电源为 _____。 Query the wiring diagram of the circuit breaker C00413, and it can be determined the power supply is _____. 飞机上跳开关 C00413 的线路分析		
4. 结束工作 **Close Out**	工作者 Perf.By	检查者 Insp.By
1）保存文件，文件用班级 + 学号 + 姓名 +WDM-03 的方式命名； 1）Save the file with the name of Class + Student Id + Name + WDM-03; 2）关机； 2）Shut down the computer; 3）6S 管理。 3）Conduct 6S Management.		

《民用航空飞机维修手册应用》实操工卡

工卡标题 Title	WDM 查询应用 4：飞机上接线盒 TB32 的线路分析 WDM Query & Application 4: Wiring Analysis of the Terminal Strip TB32 in the Aircraft			
机型 A/C Type	B737 6/7/8	难度等级 Difficulty Level		II
机号 REG.NO.	B-****	版本 Revision		R3
组别 Group		组长 Leader	组员 Team	学时 Period

参考资料 Reference	线路图手册 Wiring Diagram Manual 零部件图解目录手册 Illustrated Parts Catalog 飞机维护手册 Aircraft Maintenance Manual
警告 Warning	确保在所有起落架上都安装了锁销。没有锁销，起落架可能缩回并造成人身伤害和设备损坏。 MAKE SURE THE DOWNLOCK PINS ARE INSTALLED ON ALL THE LANDING GEAR. WITHOUT THE DOWNLOCK PINS, THE LANDING GEAR COULD RETRACT AND CAUSE INJURIES TO PERSONS AND DAMAGE TO EQUIPMENT.

编写人 Author		审核人 Reviewer		批准人 Approver	
编写时间 Author Date		审核时间 Review Date		批准时间 Approve Date	

工具 / 设备 / 材料 Tool/Equipment/Material				工作者 Perf.By	检查者 Insp.By
名称 Name	规格型号 Specification	数量 Quantity	使用情况 Usage		
计算机 Computer	Windows 7	1			
翻译软件 Translation App	有道等 Youdao etc.	1			
阅读器软件 Reader App	PDF 阅读器 PDF Reader App	1			

1. 工作任务 Requirement	工作者 Perf.By	检查者 Insp.By
飞机上接线盒 TB32 的线路分析 Wiring Analysis of the Terminal Strip TB32 in the Aircraft		

2. 工作准备 Job Set-up	工作者 Perf.By	检查者 Insp.By

1）准备好相关设备及软件。
1）Prepare equipment and software.
（1）Windows 7 及以上系统计算机一台；
（1）A computer with Windows 7 system or above;
（2）在计算机上安装有道 App 及 PDF 阅读器；
（2）Install Youdao App and PDF reader on the computer;
（3）在手机上安装有道 App。
（3）Install Youdao App on your mobile phone.

《民用航空飞机维修手册应用》实操工卡

	工作者 Perf.By	检查者 Insp.By
2）选择有效的技术文件和手册。 2）Select effective technical documents and manuals.		
3）确认注意事项和警告的相关内容。 3）Confirm caution and warning.		

3. 工作步骤 Procedure	工作者 Perf.By	检查者 Insp.By
查询飞机 B-**** 在图纸 23-22-11 上接线盒 TB32 的件号、线路连接等信息 Search for the Part Number, Line Connection and Other Information of the Terminal TB32 on Drawing 23-22-11 of Airplane B-**** * 请在手册中截出相应的图片内容说明每一步骤的查询结果 * *Please Use the Corresponding Screenshot in the Manual to Illustrate the Query Results of Each Step. * 查询步骤： Query Steps: 1. 查询飞机有效性 Check Aircraft Effectiveness 飞机 B-**** 的有效性代码（Effective Code）为 _____。 The effective code of B-**** is_____.		
2. 查询飞机在图纸 23-22-11 上接线盒 TB32（TB0032）的件号 Search the Part Number of Terminal Strip TB32 (TB0032) on Drawing 23-22-11 of the Aircraft 查询 WDM23-22-11，查询接线盒 TB32，其中 YA3 代表的含义是 _____。 查询终端清单，可知该类型的接线盒 TB32 的件号为 _____。 Search the WDM23-22-11, and query the terminal strip TB32, where YA3 stands for _____. Query the terminal strip list, you can see that the part number of this type of terminal strip TB32 is _____. 		
3. 确定在图纸 23-22-11 上接线盒 TB32 的线路连接信息 Determine the Wiring Connection Information of Terminal Strip TB32 on Drawing 23-22-11 查询 WDM23-22-11 线路图纸，可确定接线盒 TB32 一共连接了 _____ 根导线，分别为 _____、_____、_____、_____。（注意：导线的正确书写格式为 W****-****-***） 飞机上接线盒 TB32 的线路分析		

班级 Class	工作卡号 Work Card No	WDM-04	共 3 页　第 3 页

Check the WDM23-22-11 wiring drawing, you can confirm that the number of wires connected to the terminal strip TB32 is _____, These wires are _____, _____, _____, _____. (Note: The correct writing format of the wire is W****-****-***)		

4. 结束工作 Close Out	工作者 Perf.By	检查者 Insp.By
1）保存文件，文件用班级＋学号＋姓名＋WDM-04 的方式命名； 1）Save the file with the name of Class + Student ID + Name + WDM-04; 2）关机； 2）Shut down the computer; 3）6S 管理。 3）Conduct 6S Management.		

随手笔记

随手笔记

《民用航空飞机维修手册应用》实操工卡

工卡标题 Title	WDM 查询应用 5：飞机 APU 防火检测线路故障的排除 WDM Query & Application 5: Trouble Shooting of Aircraft APU Fire Dection Wiring

机型 A/C Type	B737 6/7/8	难度等级 Difficulty Level	I
机号 REG.NO.	B-****	版本 Revision	R3

组别 Group		组长 Leader		组员 Team		学时 Period	

参考资料 Reference	线路图手册 Wiring Diagram Manual 飞机维护手册 Aircraft Maintenance Manual
注意事项 Note	在 APU 火警检测器回路出现故障或跳开关 6A23 断开的情况下，执行 Fire/OVHT 测试时，MASTER CAUTION、FIRE WARN、OVHT/DET 指示灯将点亮，火警钟将响起，并且 APU DET INOP 指示灯将保持点亮状态。 WHEN PERFORMING THE FIRE/OVHT TEST WITH A FAULTED APU FIRE DETECTOR LOOP OR WITH CIRCUIT BREAKER 6A23 OPENED, THE MASTER CAUTION, FIRE WARN, OVHT/DET LIGHTS WILL COME ON, THE FIRE BELL WILL SOUND AND THE APU DET INOP LIGHT WILL STAY ON.

编写人 Author		审核人 Reviewer		批准人 Approver	
编写时间 Author Date		审核时间 Review Date		批准时间 Approve Date	

工具 / 设备 / 材料 Tool/Equipment/Material				工作者 Perf.By	检查者 Insp.By
名称 Name	规格型号 Specification	数量 Quantity	使用情况 Usage		
计算机 Computer	Windows 7	1			
翻译软件 Translation App	有道等 Youdao etc.	1			
阅读器软件 Reader App	PDF 阅读器 PDF Reader App	1			

1. 工作任务 Requirement	工作者 Perf.By	检查者 Insp.By
飞机 APU 防火检测线路故障的排除 Trouble Shooting of Aircraft APU Fire Dection Wiring		

2. 工作准备 Job Set-up	工作者 Perf.By	检查者 Insp.By

1）准备好相关设备及软件。
1）Prepare equipment and software.
（1）Windows 7 及以上系统计算机一台；
（1）A computer with Windows 7 system or above;
（2）在计算机上安装有道 App 及 PDF 阅读器；
（2）Install Youdao App and PDF reader on the computer;

《民用航空飞机维修手册应用》实操工卡

（3）在手机上安装有道 App。
（3）Install Youdao App on your mobile phone.

2）选择有效的技术文件和手册。
2）Select effective technical documents and manuals.

3）确认注意事项和警告的相关内容。
3）Confirm caution and warning.

3. 工作步骤 Procedure	工作者 Perf.By	检查者 Insp.By
飞机 B-****APU 防火测试过程中，发现故障现象为"指示正常，但喇叭不响"原因为与 APU 火警蜂鸣器相连接的插头故障，需查询连接插头的设备号、件号及线路连接信息。 During the aircraft B-****APU fire test, it was found that the fault phenomenon was "the indication is normal, but the horn does not sound". The cause of the failure is the failure of the plug connected to the APU fire warning horn. You need to query the equipment number \part number and line connection information of the connected plug. * 请在手册中截出相应的图片内容说明每一步骤的查询结果 * *Please Use the Corresponding Screenshot in the Manual to Illustrate the Query Results of Each Step.* 查询步骤： Query Steps： 1. 查询飞机有效性 Check Aircraft Effectiveness 飞机 B-**** 的有效性代码（Effective Code）为 _____。 The effective code of B-**** is_____.		
2. 查询飞机 APU 防火检测线路图 Search for the Wiring Diagram of APU Fire Detection WDM 的防火系统位于 _____ 章，发动机防火检测位于 _____ 节，APU 防火检测位于 _____ 目。 The Chapter of fire protection system in WDM is _____, the section of engine fire detection is _____, and the subject of APU fire detection is _____.		
3. 查询 APU 火警蜂鸣器相关线路 Search for the Wiring of APU Fire Warning Horn 在飞机 APU 防火检测线路图中，查询 APU 火警蜂鸣器相关线路。 In the wiring diagram of the APU fire detection, find the relevant wiring of the APU fire warning horn.		
4. 确认 APU 火警蜂鸣器（Fire Warning Horn）的线路连接信息及插头信息 Confirm Line Connection Information and Plug Information of the APU Fire Warning Horn 分析 APU 火警蜂鸣器相关线路，确认与 APU 火警蜂鸣器连接的插头设备号为 _____，其中插头的 _____ 号插钉连接的导线为 _____，插头的 _____ 号插钉连接的导线为 _____。 Analyze the wiring of the APU fire warning horn , and confirm that the equipment number of the plug connected to the APU fire warning horn is _____, where the wire connected to the _____ pin of the plug is _____, the wire connected to the _____ pin of the plug is _____.		

班级 Class	工作卡号 Work Card No	WDM-05	共 3 页　第 3 页

. 确定 APU 火警蜂鸣器连接的插头件号 Confirm Part Number of the Plug Connected to the APU Fire Warning Horn 查询 WDM 的设备清单，可确定 APU 火警蜂鸣器连接的插头件号为_____。 Query the WDM equipment list, to confirm that the part number of the plug connected to the APU fire warning horn is _____.	飞机 APU 防火测试 线路故障的排除		
4. 结束工作 **Close Out**		工作者 Perf.By	检查者 Insp.By
1）保存文件，文件用班级 + 学号 + 姓名 +WDM-05 的方式命名； 1）Save the file with the name of Class + Student ID + Name + WDM-05; 2）关机； 2）Shut down the computer; 3）6S 管理。 3）Conduct 6S Management.			

随手笔记

随手笔记

《民用航空飞机维修手册应用》实操工卡

班级 Class	工作卡号 Work Card No	WDM-06	共 3 页　第 1 页

工卡标题 Title	WDM 查询应用 6：气象雷达收发组件的故障修理 WDM Query & Application 6: Fault Repair of Weather Radar Receiver/Transmitter

机型 A/C Type	B737 6/7/8	难度等级 Difficulty Level	I
机号 REG.NO.	B-****	版本 Revision	R3

组别 Group		组长 Leader		组员 Team		学时 Period	1

参考资料 Reference	线路图手册 Wiring Diagram Manual 飞机维护手册 Aircraft Maintenance Manual
注意事项 Note	请勿触摸收发组件的插针或其他导体。如果您触摸这些导体，静电释放可能会造成收发机损坏。 DO NOT TOUCH THE CONNECTOR PINS OR OTHER CONDUCTORS ON THE RECEIVER/TRANSMITTER. IF YOU TOUCH THESE CONDUCTORS, ELECTROSTATIC DISCHARGE CAN CAUSE DAMAGE TO THE RECEIVER/TRANSMITTER.

编写人 Author		审核人 Reviewer		批准人 Approver	
编写时间 Author Date		审核时间 Review Date		批准时间 Approve Date	

工具 / 设备 / 材料 Tool/Equipment/Material				工作者 Perf.By	检查者 Insp.By
名称 Name	规格型号 Specification	数量 Quantity	使用情况 Usage		
计算机 Computer	Windows 7	1			
翻译软件 Translation App	有道等 Youdao etc.	1			
阅读器软件 Reader App	PDF 阅读器 PDF Reader App	1			

1. 工作任务 Requirement	工作者 Perf.By	检查者 Insp.By
气象雷达收发组件的故障修理 Fault Repair of Weather Radar Receiver/Transmitter		

2. 工作准备 Job Set-up	工作者 Perf.By	检查者 Insp.By

1）准备好相关设备及软件。

1) Prepare equipment and software.

（1）Windows 7 及以上系统计算机一台；

（1）A computer with Windows 7 system or above;

（2）在计算机上安装有道 App 及 PDF 阅读器；

（2）Install Youdao App and PDF reader on the computer;

（3）在手机上安装有道 App。

（3）Install Youdao App on your mobile phone.

《民用航空飞机维修手册应用》实操工卡

班级 Class	工作卡号 Work Card No	WDM-06	共 3 页　第 2 页

	工作者 Perf.By	检查者 Insp.By
2）选择有效的技术文件和手册。 2）Select effective technical documents and manuals.		
3）确认注意事项和警告的相关内容。 3）Confirm caution and warning.		
3. 工作步骤 **Procedure**	**工作者** **Perf.By**	**检查者** **Insp.By**
飞机 B-**** 气象雷达收发组件故障，需查询该组件的件号、厂商代码、位置及所在的线路图号 If the Weather Radar Receiver / Transmitter of the Aircraft B-**** Fails, You Need to Query the Part Number, Vendor, Station and the Chapter-Section-Subject of the Wiring Diagram * 请在手册中截出相应的图片内容说明每一步骤的查询结果 * *Please Use the Corresponding Screenshot in the Manual to Illustrate the Query Results of Each Step. * 查询步骤： Query Steps: 1. 查询飞机有效性 Check Aircraft Effectiveness 飞机 B-**** 的有效性代码（Effective Code）为 _____。 The effective code of B-**** is_____.		
2. 查询飞机气象雷达系统线路图章节号 Search for the Chapter-Section-Subject of the Wiring Diagram about the Weather Radar System 查询 WDM 的导航系统 _____ 章，气象雷达系统 _____ 节，气象雷达系统 _____ 目。 Inquire about Chapter _____ of WDM navigation system, section _____ of Weather Radar System, and subject_____ of Weather Radar System of WDM.		
3. 查询飞机气象雷达系统相关线路 Search for the Wiring of Weather Radar System 在飞机气象雷达系统线路图中，查询飞机气象雷达收发（XMTR-RCVR）组件相关线路。 In the wiring diagram of the aircraft weather radar system, search for the relevant wiring of the aircraft weather radar receiver / transmitter.		
4. 确认飞机气象雷达收发组件的设备号、站位等信息 Confirm The Equipment Number, Station Position and Other Information of the Aircraft Weather Radar Receiver / Transmitter 查询飞机气象雷达收发（XMTR-RCVR）组件，可确认该组件的设备号为 _____，站位为 _____。 Search for the aircraft weather radar receiver / transmitter, you can confirm that the component's equipment number is _____ and the station is _____.		
5. 确定飞机气象雷达收发组件的件号及厂商代码 Confirm the Part Number and Vendor of the Aircraft Weather Radar Receiver / Transmitter 打开 WDM 的设备清单，飞机气象雷达收发组件的件号为 _____，厂商代码 _____。 From the equipment list of WDM, the part number of the aircraft weather radar receiver / transmitter part number is_____, and the vendor code is_____. 气象雷达收发组件 的故障修理		

《民用航空飞机维修手册应用》实操工卡

班级 Class	工作卡号 Work Card No	WDM-06	共3页 第3页	
飞机气象雷达收发组件故障可按查询结果换件处理 The failure of aircraft weather racar transceiver components can be replaced according to the query results				
4. 结束工作 Close Out			工作者 Perf.By	检查者 Insp.By
）保存文件，文件用班级＋学号＋姓名＋WDM-06 的方式命名； ）Save the file with the name of Class + Student ID + Name + WDM-06; ）关机； ）Shut down the computer; ）6S 管理。 ）Conduct 6S Management.				

随手笔记

随手笔记